THE BIRTH OF THE MIND

ALSO BY GARY MARCUS

*The Algebraic Mind: Integrating Connectionism and
Cognitive Science*

THE BIRTH OF THE MIND

How a Tiny Number of Genes Creates the Complexities of Human Thought

GARY MARCUS

A Member of the Perseus Books Group
New York

Published by Basic Books,
A Member of the Perseus Books Group

Library of Congress Cataloging-in-Publication Data

Marcus, Gary F. (Gary Fred)
   The birth of the mind : how a tiny number of genes creates the complexities of human thought /
Gary F. Marcus.
      p.   cm.
   Includes bibliographical references and index.
   ISBN 0-465-04405-0
   1. Genetic psychology.   2. Psychobiology.   3. Nature and nurture.   4. Cognitive science.   I. Title.

BF701.M32   2004
155.7—dc21

2003012545

Set in 12-point AGaramond

03 04 05 / 10 9 8 7 6 5 4 3 2 1

CONTENTS

ACKNOWLEDGMENTS

THE PHILOSOPHER Peter Singer has written about an "Expanding Circle" of social contacts in which humans, over time, have learned to extend ethical concern to ever-widening groups of people—to their families, to their tribes, to other humans, to animals, and to other living things. This book is the product of a different sort of expanding circle, the expanding circle of a scientist striving to bridge a wide range of disciplines. It could not have been written without the help of my own inner circle, the greater circle of my immediate contacts, and the still greater circle of e-mail-savvy scientists whom I've never even met but who took the time to answer my questions on topics ranging from gene-counting to Princess Leia's career prospects.

Four people deserve special mention.

Simon Fisher read every word and made me reconsider each and every one; his Talmudic commentary on the penultimate draft was nearly as long as the manuscript itself (or maybe it only seemed that way!), a master class from a new and already treasured friend.

Athena Vouloumanos read not just every word but nearly every draft of every word; she was with me from the first word of the first outline to the last word of the index. Never one to give praise where it was undeserved, she pushed me to rewrite and rewrite until she liked every line. She may be named after a goddess, but she was my Muse.

Steve Pinker, too, read every word, and before that, in the early 1990s, gave me the training that stimulated me to think about the very questions that animate this book. It was a true privilege to learn to write—and to think—while watching over his shoulder; as an apprentice I couldn't have asked for more.

My father, Phil Marcus, was my first teacher. His probing, penetrating intellect—and his insistence that I always look beyond the here and now of my own research to how it all fits together—has its stamp on every page.

Over a dozen more, with training in fields from cognitive development to evolution, medicine, behavioral genetics, and biochemistry—I hope I haven't missed anyone—read drafts and made detailed comments, catching howlers, calling me on obscurities, and educating me in the finer details of their own disciplines: Yuri Arshavsky, Shoba Bandi Rao, Derek Bickerton, Luca Bonatti, Rob Boyd, Tom Clandinin, Barbara Finlay, Peter Gordon, George Hadjipavlou, Ellen Markman, Liqun Luo, Franck Ramus, Rasmus Storjohann, Heather van der Lely, and Essi Viding.

Others, including Evan Balaban, Iris Berent, Istvan Bodnar, Beth Bromley, Anders Ericsson, Judit Gervain, Kristen Hawkes, Ray Jackendoff, Kathleen Much, Ivan Sag, Dan Sanes, Ralf Schoepfer, Mark Turner, Barbara Tversky, and Fei Xu, read and commented on selected chapters.

Audiences at dozens of universities, including Arizona, Berkeley, CMU, Cornell, CUNY, Oxford, Rochester, Stanford, UBC, UC Davis, UCL, UCLA, the University of Toronto, and Yale, and a terrific set of workshops on innateness organized by Peter Carruthers, Steve Stich, and Steven Lawrence, helped sharpen my thinking. Bruce Baker, Noam Chomsky, Michael Cooke, Sarah Dunlop, Douglas Frost, Randy Gallistel, Lila Gleitman, Justin Halberda, Marc Hauser, Jennifer Lee, Kathy Nordeen, Laszlo Patthy, Samuel Pfaff, Sandor Pongor, Todd Preuss, Rob Sampson, Greg Sutcliffe, Michael Tomasello, Ajit Varki, Michael Weliky, Al Yonas, and the crew at the ensembl.org helpdesk patiently answered my queries.

New York University gave me two years' leave; Keith Fernandes kept the lab going while I wrote. The Center for Advanced Study in Behavioral Sciences at Stanford sheltered me in the rare moments of California rain and brought me together with Mark Turner and a wonderful crew of fellow Fellows, especially Ivan Sag, Anders Ericsson, Istvan Bodnar, Bob Brandom, Malka Rappaport Hovav, Danielle

MacBeth, and Kristen Hawkes, not to mention the many superb linguists, psychologists, computer scientists, and biologists down the hill.

Financial support came from the National Institutes of Health (NIH), the Human Frontier Science Program (HFSP), and the MacArthur Foundation.

Katinka Matson helped me find a home for my manuscript. Jo Ann Miller guided me on the art of writing for the general public; Kathy Streckfus polished the manuscript to a shine; Tim Fedak of figs.ca helped illustrate it; and, Rich Lane and Felicity Tucker and their team turned a manuscript into a beautiful book.

And finally, I thank my mother, Marilyn Marcus; my stepmother, Linda; my sister, Julie, and my Aunt Esther and Uncle Ted, for always encouraging me in whatever I've endeavored to do.

Thank you all! One more sentence and you'd all be coauthors!

1

NEITHER IS BETTER

The genetic code is not a blueprint for assembling a body from a set of bits; it is more like a recipe for baking one from a set of ingredients. If we follow a particular recipe, word for word, in a cookery book, what finally emerges from the oven is a cake. We cannot now break the cake into its component crumbs and say: this crumb corresponds to the first word in the recipe; this crumb corresponds to the second word in the recipe, etc.

—Richard Dawkins

FRANCIS CRICK, codiscoverer of the structure of DNA, recently argued in his book *The Astonishing Hypothesis* that the activity in our minds has its basis in our brains: "To understand ourselves, we must understand how nerve cells behave and how they interact."[1]

Crick is surely right that the mind arises from the activity of the brain. But, having grown up in the late twentieth century, the son of a software engineer who once studied the biophysics of neurons, I can't say that I am astonished. To many people of my generation, it has become obvious (maybe even banal) that our thoughts are the product of our brains. In the words of MIT cognitive scientist Steven Pinker, "The mind is what the brain does."[2]

In contemporary society, we are surrounded by evidence of the influence of the brain on the mind. Science has shown that Prozac affects our mood by targeting the brain, that strokes can cause brain lesions that alter our behavior, and that distinct parts of the brain are active in different aspects of cognitive functioning—the right brain when we listen to music, the left when we listen to speech,[3] the amygdala when we are frightened,[4] the right prefrontal cortex during orgasm.[5]

But although most people have by now accepted the fact that the mind has its origins in the brain, far fewer have become comfortable with a second fact: that the origins of the brain are in the genes. The molecule that Crick helped to decipher just over fifty years ago has had an enormous impact on science, medicine, even law. Yet it has had almost no impact on theories of the mind.

If genes can predispose us to cancer or diabetes, it stands to reason that they might significantly shape our minds. It is easy to admit that genes have something to do with why one breed of dogs is friendlier (or meaner) than another, but even scientists can be reluctant to accept the notion that they might affect our own thoughts and behavior. In a recent issue of *Current Anthropologist,* two Stanford biologists, Paul Ehrlich and Marcus Feldman, wrote that "the concept of overall heritability should be restricted in its employment to plant and animal breeding. . . . [When it comes to humans] genes can control some general patterns . . . but they cannot be controlling our individual behavioral choices."[6]

Ehrlich has gone so far as to argue that the effect of genes must be limited because of what he has dubbed a "gene shortage." Our species has perhaps 30,000 genes, yet our brains have on the order of 20 billion neurons. "Given that ratio," Ehrlich concluded, "it would be quite a trick for genes typically to control more than the most general aspects of human behavior."[7] This view was recently echoed in the writings of cultural critic Louis Menand, who, in the pages of *The New Yorker,* wrote that "every aspect of life has a biological foundation in exactly the same sense, which is that unless it was biologically possible it wouldn't exist. After that, it's up for grabs"[8]—echoing an old boast by John B. Watson (no relation to Crick's collaborator James) that he could raise any child to do anything, so long as he had his "own specified world to bring them up in."[9] People don't want to accept that genes play an important role in our mental life because this notion challenges our sense of being able to shape our own destinies.

But it is patently clear that genes do shape our mental lives. Although Ehrlich and Feldman are, strictly speaking, correct—genes certainly don't *control* our destinies—genes do contribute to our personalities, our temperaments, and the qualities that make each individual unique, as well as to the qualities that make the human species unique. Modern science has revealed dozens of ways in which genes have a demonstrable effect on mental life. Animal studies have shown that aspects of behavior and personality can be genetically transmitted (as in the example of the dog breeds that I mentioned earlier, and in studies in which mouse geneticists have bred rodents to be as anxious as Woody Allen).[10] Studies of twins have shown time and again that people who share more genes (such as identical twins) are more similar than those who share fewer genes (such as fraternal twins), not only in physical attributes, but in personality and intelligence, indeed just about anything mental that can be measured.

Of course, not every similarity between twins depends on genes. When two seventy-one-year-old Finnish twins died within hours of each other on March 6, 2002, each of a bicycle accident, each while riding on the same road, it really was just a coincidence.[11] Genes might have predisposed them both to enjoy physical activity or to enjoy taking risks, but it was sheer chance that caused them to die on the same day and in the same way. Nevertheless, the influence of genes on our mental structure is undeniable.

This influence extends to the structure of the brain itself. For example, a team of brain imagers from the University of California at Los Angeles combined with a team of geneticists from Helsinki to take three-dimensional magnetic resonance images of the brains of twenty sets of twins—ten identical, ten fraternal—carefully matched in terms of their social class, when they were born, and how much time they had spent together.[12] The density of the gray matter of the brain—the part of the brain that is most likely to stay constant regardless of experience—was much more similar in the brains of identical twins than in the brains of fraternal twins.

Another team found that the volume of *white* matter—the part of the brain that consists of modifiable neural connections and that might be expected to be *most* influenced by experience—was also more similar in identical twins.[13] The brains of identical twins are more similar than those of fraternal twins in the patterns of convolutions[14] and in the size of particular structures, such as the corpus callosum (which connects the left and right hemispheres).[15] Studies with cats suggest that those similarities may extend even to finer-grained details, such as the spacing and layout of microscopic cortical columns, sets of densely connected brain cells that share functional properties.[16] Genes thus appear to shape even the finest details of the brain.

Yet another hint that genes must play an important role in the development of the mind comes from newborn babies. Within hours of their birth, newborns can imitate facial gestures,[17] connect what they hear with what they see,[18] distinguish the rhythms of Dutch from the rhythms of Japanese,[19] and tell the difference between someone who is looking at them and someone who isn't,[20] suggesting that even with relatively little experience, newborns are ready to start observing the world. Building on the ideas of the pioneering linguist Noam Chomsky, "nativists" such as Steven Pinker and the French cognitive neuroscientist Stanislas Dehaene have argued that babies are born with a "language instinct"[21] and a built-in "number sense."[22] The tradition of a newborn as a "blank slate" shaped solely by experience (uninfluenced by genes) is, as Pinker has forcefully argued, no longer tenable.[23]

∞

By now, these results shouldn't come as news. But whether we read newspapers and magazines or the professional literature in psychology, we find very few theories of the mind that make genuine contact with genes; in psychology, it's almost as if Watson and Crick never met DNA.

My goal in this book is not to try to prove that genes make a difference—a matter that is no longer in serious doubt—but to describe *how* they work and to explain, for the first time, what that means for the mind. I won't argue that genes dictate our destinies (they most certainly do not, and I'll explain why not), nor that they outweigh the contributions of culture or experience (which are difficult to measure). The thesis of this book is that the only way to understand what nature brings to the table is to take a look at what genes actually do.

Almost everything that is written about genes in the popular press is misleading in one way or another. We read that genes are blueprints or maps. We are told that they are like books, libraries, recipes, computer programs, codes, or factories.[24] But we're never let in on the secret of what genes really do. Thus, readers have no basis with which to evaluate competing claims. Could evolution have built a language instinct? Is there truly a gene shortage? Without a clear explanation of how genes work, there is no way to tell. What does it mean when newspapers report that a gene for alcoholism or obesity has been discovered? There is no way to interpret the daily onslaught of exciting biological discoveries without understanding what genes actually do.

In order to understand how genes influence human traits and capacities, we must first abandon the familiar idea of a genome (the set of genes within a particular organism) as a blueprint. The genome is not an exact wiring diagram for the mind or a *picture* of a finished product, even if newspaper headlines so often seem to suggest otherwise. Athena was said to have sprung fully formed from the head of Zeus, and the seventeenth-century scientists known as "preformationists" thought that babies were tiny, fully formed creatures within the sperm or egg cells in which they originated. But nowadays, biologists realize that in early development, such little creatures are not to be found. There are at least five good reasons to think that genomes do not provide detailed blueprints that specify a final product in intricate detail:

- In blueprints, there is a direct correspondence between the elements of the drawing and the elements of the building it describes. There is no such one-to-one correspondence between genes and the cells and structures that make up an organism. As British zoologist Patrick Bateson put it, "The idea that genes might be likened to the blueprint of a building . . . is hopelessly misleading because the correspondences between plan and product are not to be found. In a blueprint, the mapping works both ways. Starting from a finished house, the room can be found on the blueprint, just as the room's position is determined by the blueprint. This straightforward mapping is not true for genes and behaviour, in either direction."[25]

- A blueprint that is 1 percent different from the next yields a building that is 1 percent different. But a genome that is 1 percent different can lead to a radically different mind. A single change in our genetic makeup can lead to disorders ranging from sickle-cell anemia to certain kinds of specific language impairment. Our genomes are only about 1 percent different from those of chimpanzees, yet our minds are radically different.

- Genomes are too small to contain the kind of detail one would expect if genes were truly an exact blueprint for the wiring of the mind. The human genome contains far fewer than 100,000 genes—perhaps as few as 30,000,[26] paltry in comparison to the 20 billion or so neurons found in the human brain.[27] Ehrlich's gene shortage militates against any idea of the genome as a literal blueprint.

- Identical genomes do not yield identical nervous systems. In the mid-1970s, neurobiologist Corey Goodman showed that the nervous systems of grasshopper clones with identical genotypes were similar, but not identical.[28] More recent studies using newly developed brain imaging technologies have shown that the same is true of human twins: The brains of identical twins are similar, but decidedly not identical.[29]

- Just as identical twins do not have identical brains, they also do not have identical minds. One twin may be more ambitious, the other more nurturing. These differences, presumably, correlate with differences in brain structure. Identical twins can differ in weight, religion, and even sexual orientation. Even with identical genomes, identical twins are separate people with separate minds.

Clearly, the blueprint metaphor is flawed. Yet, as we will see, many discussions of nature and nurture founder precisely because they wrongly assume genes to be simplistic blueprints.

The second biggest misconception people harbor about genetics: that it will be possible one day to determine, once and for all, whether nurture or nature is "more important." Genes are useless without an environment, and no organism could make any use of the environment at all if it were not for its genes. Asking which one is more important is like asking which gender, male or female, is more important, as the deliberately obtuse British comedian Ali G did in an interview with a feminist scholar. (I quote the dialogue verbatim, in Ali G's own unique dialect. Fans of his will recognize that Ali G is a fictional character and that the whole bit is shtick. True fans will know that the talented young man who plays Ali G, Sascha Baron-Cohen, is a cousin of Simon Baron-Cohen, one of the world's leading scholars of cognitive development.)

Ali G: Now one in two people in the country is a woman. We's got to know about them. Women: They is important aren't they?
Professor: They indeed are. Very important. As important as men.
Ali G: Which is better? Man or woman?
Professor: Well, equality is not about being better.
Ali G: But which one is better?
Professor: Either is better.
Ali G: But one must be a little bit better . . .
Professor: [pauses] In what respect?
Ali G: Like in the way, you know, that something is worse and something is better.[30]

In the interaction between nature and nurture, neither is better. The better question is not "which" but "how": How do genes work together with the environment to build a human mind?

Before we begin to tackle this question, let me note that there actually is a statistic known as "heritability" that at first blush seems to measure which one, genes or environment, is "better." More precisely, heritability is an estimate of the contribution of genes and the environment to *individual differences* in any given trait. Are differences in intelligence between people more a matter of genes or the environment? How about assertiveness? Neuroticism? Self-discipline?

To answer these questions, researchers can assess and calculate how individual differences in attributes such as IQ or personality vary as a function of genetic relatedness.[31] Heritability is determined not by poring over DNA sequences (no microscope required) but by comparing the total amount of variation in one trait with the extent to which that variation is shared between related people. If, controlling for environment, closely related people are significantly more similar on a particular trait than less closely related people, that trait is said to be highly heritable.[32] As you might expect, fingerprints come as close as anything to being set in stone (nature), whereas the extent to which hands are callused is largely a function of one's line of work (nurture). Some physical traits, such as biceps size, are a mix of an individual's inherent constitution and experiential factors such as diet and workout regime. Similarly, most mental traits fall somewhere in the middle. For example, "identical" twins (who share all their genes) are never actually identical, but on almost anything one can measure, they are more similar than fraternal twins (who share only half their genes): IQ, temperament, even the extent to which they are religiously devout.[33] Likewise, siblings are more similar than half-sibs or cousins.[34]

Heritability scores, in principle, can range from 0 percent, which would mean that none of the differences between individuals can be attributed to differences in genes, to 100 percent, which would mean that all the differences between individuals can be attributed to differences in genes. The heritability of getting hit by lightning would come in at close to zero—in other words, getting hit by lightning is not at all determined by genes. Fingerprints, in contrast, come in at nearly 100 percent—individual differences in fingerprints are almost completely genetically determined.[35] In nearly every measure of the mind, scores are well above 30 percent, and often as high as 60 to 70 percent. That's high enough that we can be confident that genes are in some way involved, but low enough to make it clear that there is something beyond genes (it could be environment, or it could be random chance) that is important.[36]

Heritability scores have an air of authority, but they are easily misunderstood. For example, it is tempting to interpret a heritability score of 60 percent on IQ tests as showing that "60 percent of intelligence comes from heredity."[37] Although twin studies do suggest that IQ has a heritability that is not far from 60 percent, that does not mean that 60 percent of your intelligence comes from your genes. In fact, the heritability measure doesn't say what percentage of any trait comes from the genes. Here's why.[38]

First, heritability doesn't reveal what percentage of a trait comes from the genes, it only measures what percentage *of the variation in that trait* can be attributed to those genes. What do I mean by "percentage of variation in that trait"? (I'll get to the equally tricky phrase "attributed to those genes" in the next paragraph.) Heritability measures can't see the forest for the trees; all they can see is the differences between trees. What enters into the statistic is not the average height of the trees, but the *differences* in height between them. As a consequence, heritability can only speak to what makes some trees bigger than others (is it light and moisture, or just rapid-growth genes?), not what makes a tree have a trunk or roots. In humans, heritability only looks at differences that in the grand perspective of life on earth are tiny: whether Jimmy has a bigger vocabulary than Johnny or whether Janey is better with a wrench than Susan, not that which makes all humans intelligent creatures. It is entirely possible that 5,000 different genes contribute to human intelligence and that only a few hundred of those vary in ways that contribute to the differences between one person and the next. In the words of psychoanalyst Harry Stack Sullivan, oft-repeated by my mother, a social worker and lifelong student of human nature, "We are all more human than otherwise." Heritability scores tell us only how differences in those few genes correlate with differences in scores like IQ, not about the contribution of the genes that we all share, or of how genes contribute to making humans different from chimpanzees.

Second, saying that a trait can be "attributed" to genes is not the same thing as saying it is *caused* by genes; heritabilities are just measures of correlation, and correlation never guarantees causation. Almost all Jedi Knights are male and hence bear Y chromosomes, so statistically speaking, the chance of being a Jedi Knight is tied to the presence or absence of a Y chromosome. But Princess Leia may have the Force, too; perhaps the real problem is not a lack of talent, but a lack of opportunity—maybe the Jedi powers-that-be in her era tended not to give females equal consideration for Yoda's Jedi boot camp (though I hear that equal opportunity could reach the Force in Episode VII).[39] Y chromosomes would still then be correlated with who gets to be a Jedi, but they would not be a *cause* of being one. Likewise, differences on verbal IQ tests can be statistically related to the genes for gender, but that doesn't mean those differences are caused by those genes—they might instead result from how society treats people of different genders. By treating all relations, causal or otherwise, the same, heritability scores can mislead us about the contribution of genes to the finished product.

Third, as any behavioral geneticist would explain, heritability scores inevitably reflect the range of environments from which the data are collected.[40] IQ tests taken from a homogeneous society in which all children receive compulsory education—say, the contemporary United States—tend to minimize the possible impact of environmental variation and thereby yield relatively high heritability scores. Heritability measurements taken from a society with more radical variation—such as an earlier period in the United States when only the wealthy could afford education—would likely yield lower estimates of heritability. There simply is no fact of the matter, no absolute number. Heritability scores are a little like an auto manufacturer's gas mileage estimates. They may be a good relative guide (body weight is more sensitive to environment than fingerprints, just as your subcompact will get better mileage than my four-wheel-drive SUV), but the absolute numbers mean relatively little. Whether you get 35 or 26 miles to the gallon will depend on the road you travel, how recently your car has been tuned, how aggressively you drive, and so forth. In a similar way, any heritability score is a complex reflection of the measure used and the population studied.

Until recently, there wasn't much more to say. Scientists knew that nature and nurture both mattered, but they didn't know why or how. As the late Nobel laureate Peter Medawar put it in 1981, the only tool biologists had for investigating the development of the mind was to study differences between individuals. If we wanted to know whether "any common characteristics human beings possess" (for example, the ability to learn language) were in some way "genetically encoded and part, therefore, of our inheritance," we were stuck.[41] Such theories might be true but we had no way to verify them. The debate between nature and nurture was widely seen as a "wearisome,"[42] unanswerable question.

But much has changed since 1981, and we are finally in a position to move past the long-standing impasse, not by trying to decide which one is better, but by trying to better understand how the two— genes and the environment—work together. Newly invented biological techniques allow scientists to assess the contributions of individual genes, and even to deliberately alter those genes, launching a whole new scientific enterprise that studies the molecules that help shape the mind. The goal of this book is to unite the results of groundbreaking scientific research with studies of the psychology of humans and other animals—in other words, to take insights from the genome and use them to revamp our understanding of nature, of nurture, and of how they work together to create a human mind.

To do that, I must immerse you in a world of cells and proteins, the domain in which genes actually do their work. That may seem peculiar in a book about the mind—most books about the mind are about psychology, not cells—but my argument is that the workings of the cellular world cast enormous light on the mental world, and that the mental world cannot be properly understood without a firm grip on the cellular world. Anything else would just be business as usual, "nature and nurture" without the nature.

Any theory that puts the role of genes front and center must deal with two of the most difficult challenges in the science of the mind, which I will call the Two Paradoxes. First, any adequate theory must face the challenge of neural flexibility. For every study that tells us that a newborn can understand something about the world, there is another that shows that the brain can continue to function even when its structure is altered. How can the mind be at once so richly structured and so flexible? The second challenge is Ehrlich's "gene shortage": How can the complexity of the brain emerge from a relatively small genome, 20 billion neurons versus just 30,000 genes?[43]

∽

This book, then, is about the mind, the brain, and the molecules that make them what they are. I begin, in Chapter 2, with the mind, asking what a newborn understands (and does not understand) about the world. How is a newborn human different from a newborn chimp or a newly hatched bird? As both the culmination of nine months of intricate self-assembly (performed in the comforts of the womb) and the beginning of a lifetime of learning and experience, birth is the perfect place for us to begin our investigation. As I will argue, we are, more than anything else, born to learn.

In Chapter 3, I turn to the brain. What is the structure of a newborn's brain, and how does it relate to the structure of an adult's brain? The key focus in Chapter 3 is the paradox of flexibility, the tension between the seemingly intricate structure of a newborn's brain and the tremendous flexibility with which it develops. My conclusion is that nature bestows upon the newborn a considerably complex brain, but one that is best seen as *prewired*—flexible and subject to change—rather than *hardwired,* fixed and immutable.

Beginning with Chapter 4, I turn to genes and proteins, the primary stuff of which brains are made. Rather than painting pictures, genes provide recipes for proteins and crucial instructions for when those recipes should be made and put to use. The discovery of these genetic "recipes" is the story of how scientists came to understand the true nature of genes.

Next, I show what it is that genes bring to the problem of building brains, focusing on a simple truth that is unsurprising yet freighted with profound implications: Genes play almost exactly the same role in building the brain as they do in building any other part of the body. From a mind's-eye view, brains may seem awfully special—unlike anything else in the universe—but from a gene's-eye view, brains are just one more elaborate configuration of proteins. Chapter 5 puts the development of the human brain into the context of the rest of biology.

Chapter 6 is about what most makes the brain special—the elaborate system of "wires" that run between nerve cells, and how those neuronal wires are laid down and revised over time. My goal is twofold: first, to show how important genes are even in the process of wiring, and second, to reveal how the environment eventually gets involved in the process of building a person.

In Chapter 7, I explore the origins of the genes that contribute to the construction of the mind and try to place the human brain in its evolutionary context. In so doing, I take on the question of why humans, but not chimpanzees, are able to speak and acquire rich culture, given that our genomes are 98.5 percent similar.[44]

In Chapter 8, I show how an understanding of genes as self-regulating recipes helps to cut through the Two Paradoxes, how genes allow innateness to coexist with developmental flexibility, and how they permit intricate structures to emerge from a relatively small genome.

The final chapter ties these threads together, showing how a synthesis of biology and the cognitive sciences is leading to a new understanding of nature and nurture, and what that might mean for our future.

2

BORN TO LEARN

Everyone is born a genius.

—R. Buckminster Fuller

SCIENTISTS HAVE BEEN wondering for years what the mind of the newborn is like. One fateful July 1993 *Life* magazine article told us that "Babies Are Smarter Than You Think," and ever since then newspapers and magazines have been filled with stories of clever babies. (I'm partly responsible. One study from my lab induced an editor at the *Rock Hill Herald* in South Carolina to title a story "Scientists Go Ga-Ga over Babies' Goo-Goo."[1])

But now there's finally been a backlash. The latest headline says "Study Reveals: Babies Are Stupid." In the accompanying story, we are told that when babies were prodded with a broom handle, "over 90% of them . . . failed to make even rudimentary attempts to defend themselves. The remaining 10% responded by vacating their bowels." A photo shows a baby gnawing on a squeeze toy, with the caption "Despite their relatively large cranial capacities, babies such as this one are so unintelligent that they are unable to distinguish colorful plastic squeak toys from food sources."[2]

The study is, of course, a satire, written by the clever folks at *The Onion.* No real scientist has ever used a broom handle to study the cognitive capacities of a human infant. But behind the jokes there is a deep truth about the difficulty of doing good psychological experiments. If we discovered that most babies who were poked with a broomstick didn't move, how would we interpret this finding? Would it be because the babies didn't realize the danger? Because they trusted the experimenter not to injure them, or because they didn't have the capacity to fight back? Any time a baby "fails" in an experimental task, the careful scientist must ask why.

Noam Chomsky has made a distinction between "competence" and "performance" that serves as a handy guide for wading through the complex literature on what babies know.[3] According to Chomsky, one can distinguish between the competence, or ability, to do something in principle, and the practical obstacles to actual performance—ranging from limits on memory to limits on motor abilities—that can keep someone from putting that competence to good use. The fictional *Onion* babies might have had the in-principle competence to recognize danger without having the performance abilities to do anything about it.

Here's a real example, drawn from one of the longest sagas in contemporary developmental psychology. Pick up any developmental psychology textbook from more than a decade ago, and you will read that at eight months of age, babies do not yet realize that objects continue to exist when they are hidden from view. In the jargon of psychology, the infants lack "object permanence." The evidence for this dramatic statement came mainly from an observation by Jean Piaget, generally considered to be the father of cognitive development. Intensively studying his own children (in a tradition started by Charles Darwin), Piaget came to the conclusion that for young infants, objects that were out of sight were not simply out of mind, but out of existence. According to Piaget, for a young infant, "a vanished object is not yet . . . a permanent object which has been moved; it is an image which reenters the void as soon as it vanishes, and emerges from it for no objective reason."[4]

One of Piaget's tests was to show a child an interesting toy and then see how the child would respond when the toy was hidden under a blanket. When his daughter Lucienne was eight months old, Piaget showed her a toy stork and allowed her to hold it and shake it. Lucienne was clearly quite taken with the stork. But when Piaget hid it, Lucienne ceased paying any attention to it. Piaget wrote, "As soon as the stork disappears under the cloth, Lucienne stops looking at it and looks at my hand. She examines [Piaget's hand] with great interest but pays no more attention to the cloth."[5]

Piaget found the same thing with his other children, Laurent and Jacqueline, and many other scientists have since replicated the result. But does this mean that eight-month-old babies don't recognize the permanence of objects? A series of more recent experiments—which ease the performance demands by taking the act of reaching out of the task—suggests otherwise. One of the first, known as the "drawbridge" study, was conducted in 1985 by the psychologists Renée Baillargeon, Elizabeth Spelke, and Stanley Wasserman.

Building on an earlier observation by the psychologist Robert Fantz[6] that babies often pay more attention to things that are unexpected, they designed an experiment in which all babies had to do was watch—no reaching required. Five-month-olds watched a drawbridge-like screen raise and lower, over and over again, until they were bored. At this point, the experimenters placed a box behind the drawbridge, directly in its path, and then motion of the bridge resumed. Each time the bridge came up, the box gradually became hidden. The test was to see if babies would care whether the drawbridge would now do just what it had done before, continuing through 180 degrees of rotation (a seemingly impossible event, given that the box ought to be in the way), or if they would expect the drawbridge to stop when it reached the box (a "possible" event).

To an adult, the former "impossible" event is surprising, reminiscent of a magic trick, and the latter "possible" event is hardly worth noticing. Baillargeon and her collaborators discovered that five-month-old babies, too, were "surprised" by the impossible event, looking longer at it than they did at the "possible" event.[7] A control experiment showed that if the box was placed outside the drawbridge's path of movement, babies showed no preference. In contrast to what Piaget had found using less sophisticated methods, the experiment

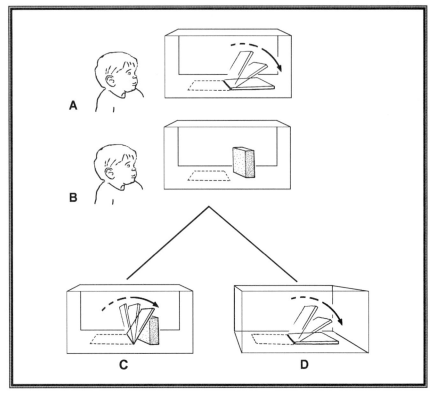

Figure 2.1   The drawbridge experiment
Illustration by Tim Fedak

suggested that the babies knew the box was there even when the draw-bridge came up to hide it.

Although one can poke holes in this particular experiment and come up with alternative hypotheses,[8] its general conclusion remains valid; dozens of subsequent experiments with different methods showed the same thing: Well before they are eight months old, babies expect a hidden object to continue to exist. For example, psychologist Karen Wynn showed that five-month-old infants could keep track of how many Mickey Mouse dolls had been put onto a puppet stage, even if some were hidden behind a screen that rose up from the stage floor.[9]

Why, then, didn't little Lucienne just pull away the blanket and get the stork she so richly deserved? It's not because she couldn't reach for the blanket. Ingenious experiments by Yuko Munakata have shown that children can do the kind of "means-[to an]-end" analyses that would be necessary: Babies can push a lever to get an object that is underneath a transparent cover.[10] My own view is that the reason an eight-month-old won't pull away the blanket has to do with how babies set their goals. Babies do not construct to-do lists. Instead, they seem to decide what to do at any given moment mainly by looking around at the "here and now" and focusing on the most interesting stimulus around. The toy is interesting. If they see it, they go for it. If the toy is covered by the blanket, they'll go for the blanket, but only if the blanket is much more interesting than anything else around. If mom is around, forget about the blanket.[11]

Before you laugh at your baby's now-you-see-it, now-you-don't distractibility, think of Steve Martin's pastime for "the over-fifty set":

> *Bored?*
> *Here's a way [you] can easily kill a good half-hour:*
> —*Place your car keys in your right hand.*
> —With your left hand, call a friend and confirm a lunch or dinner date.
> —Hang up the phone.
> — Now look for your car keys.[12]

Just because you can't see your car keys doesn't mean you don't think that they exist. Although babies are easily distracted, they are perfectly capable of keeping track of hidden objects. Put them in the dark (where there is nothing else to do instead of covering an object with a blanket, leaving a whole room of other interesting things to look at), and young babies happily reach for objects that are no longer visible to them, even adjusting their hand grip to match the size of objects they can't see.[13]

It's not just that babies are born knowing about objects. At birth, they also seem to know something about faces, words, and maybe even sentences. In 1991, British psychologists Mark Johnson and John

Morton discovered that newborns can tell the difference between a line drawing of a face and a figure in which those lines had been scrambled.[14]

Johnson and Morton put such drawings on the back of a hand mirror and moved them slowly back and forth in front of the babies' eyes. Babies spent more time following the picture of a face than they did the picture of a scrambled face. Another, even more subtle study showed that newborns also know something about eye gaze, paying more attention to those who look straight at them than to those who face them but look away.[15] If you want a baby to look at you, don't avert your eyes.

French psychologists Jacques Mehler, Thierry Nazzi, and their colleagues have shown that babies can distinguish between words with different pitch structures (low-high versus high-low) and between sentences with different rhythms in languages they had never heard before.[16] University of British Columbia psychologists Athena Vouloumanos and Janet Werker have shown that newborns already have a bias for human speech, even in comparison to similarly complex sounds that are nonlinguistic.[17]

It's not *all* there at birth. Babies aren't nearly as good as adults at recognizing faces or languages—at two days old, babies cannot distinguish between a detailed picture of a face and a simpler, fuzzier picture of a few blobs in the right orientation, or between a grammatical sentence and an ungrammatical one with the same rhythm. Instead, what developmental psychologists have learned is that children are born with sophisticated mental mechanisms (nature) that allow them to make the most of the information out there in the world (nurture).[18]

WE ARE NOT ALONE

Babies come by their talents honestly, by dint of evolutionary hard work. Nature has been growing innate "instincts" for as long as it has been growing animals at all. Newly hatched chicks, for example, seem to have object permanence[19] and can tell the difference between a bunch of dots that move in a biological way (as if they were lights attached to prominent points of a hen who was walking in a darkened room) and dots that move at random.[20] Itty-bitty Labrador puppies can track their owner's eye gaze.[21] Horses can control their muscles well enough to walk (or at least wobble) within minutes of their birth.

In some animals, considerably more complex series of behaviors seem to be built in. Consider, for example, the courtship routine of a male fruit fly, which follows a predictable sequence even if that fly has never seen the process performed. He will begin by turning toward a female; if all goes well, he will then begin to follow her. Next, he will tap her with his forelegs and, if still not rebuffed, sing her a song by vibrating his wings. He will then proceed to lick the genitalia of his intended mate. At that point, he will curl his abdomen and finally consummate the relationship.[22] Experience plays a role in how the male chooses a mate, but once the courtship process begins, the steps are largely fixed, proceeding essentially the same way every time. As we will see later, researchers have come a long way in identifying what genes are involved in this process.[23]

Grooming reflexes for many animals are often equally formulaic. A typical mouse will start with its head, proceed to its torso, move to its anal-genital region, and conclude with its tail.[24] So strong is the urge to do this that mice will try it even when their paws have been amputated from birth.[25] In the summer of 2002, scientists from the University of Utah uncovered a gene that plays a role in modulating the whole sequence.[26] Without it, mice groom themselves incessantly, to the point of tearing out their hair. The grooming process of the red junglefowl is even more elaborate. Once every couple of days, whether he needs it or not, the junglefowl indulges in a process called dust-bathing, which is far more complex than the basic lather-rinse-and-repeat described on the back of your shampoo bottle.[27] Junglefowl don't require adult models, and they don't even require dust: They will do the dustbathing dance even when they are raised in isolation on a special wire floor that prevents dust from accumulating.

Animals are born with more than these canned sets of stereotyped behavior patterns. For example, Harvard University psychologist Marc Hauser has shown that cotton-top tamarins, small monkeys from northwestern Colombia, can be trained to use tools to reach for Fruit Loops—even though they rarely use tools in the wild. The tamarins are clever enough to use canes that have the right shape in the right orientation (with the loop capable of snagging the object) and clever enough to reject canes that have the wrong shape or that are in the wrong orientation. Whenever Hauser gave the tamarins a choice between a tool that would work and one that wouldn't, the tamarins always went for the one that would bring home the snack.[28] Tamarins obviously aren't born knowing about canes, but, like many animals, they are born with powerful abilities for analyzing the world.

Analyzing the world would, however, be of relatively little use if animals couldn't remember the results of their analyses. Luckily, most, perhaps all, animals are born not just with the ability to perceive and act but also with the ability to learn and to use past experience to improve subsequent behavior. Remarkably, it does not take very sophisticated neural hardware for a creature to be able to learn a little. What ethologist Peter Marler called an "instinct to learn"[29] is found even in worms, which can learn which side of a dish has more food, and sea slugs, which can learn to ignore the irritating prods of curious experimenters.[30] In the lingo of experimental psychologists, the worms learn to *associate* food with a particular location, and the slugs *habituate* to the endless prodding.

Association and habituation are just two of the many forms of learning evident in the animal world—two of the most ubiquitous, as it turns out. Your goldfish can do them, your dog can do them, and even your not-so-swift cousin Joey can do them. When John Watson said he could make any child a doctor, a lawyer, a beggerman, or a thief, he had similar kinds of techniques in mind.[31]

Habituation and association have their uses, but as psychologist Randy Gallistel has noted, the animal world is filled with far more interesting talents for learning, talents that apply sophisticated cognitive analyses to problems of learning highly specific information.[32] Consider, for example, the mechanism by which a bird known as the indigo bunting learns about the night sky. Why, one might ask, would the bunting care about the sky? Because the bunting wants to know which way is south. Like many of its well-heeled human counterparts, it summers in the eastern United States and winters in the Bahamas. To get to there from here, the bunting uses the stars as a navigational guide. Rather than simply memorizing that Polaris marks north, buntings actually orient themselves by watching how the stars rotate.[33]

Since the stars rotate only 15 degrees per hour, orienting by watching the heavens rotate is like watching paint dry. But the bunting perseveres, and it winds up with a far more robust tool for navigation than if it had only learned where a particular star is. The bunting is not bothered by a cloud here or there—it doesn't need to know where Polaris in particular is—and the same system will work even as the earth's place with respect to the heavens changes. The bunting's built-in celestial learning mechanism is much handier than any soon-to-be-out-of-date edition of *Hammond's Star Atlas* could ever be. What this shows is that the bunting's navigation system is a mix of something built in (a system for calibrating built-in mechanisms to local conditions) and something learned (the particular local conditions).[34]

Honey bees, too, use a highly specialized learning mechanism to help them figure out where they are going; the difference is that their system works based on the trajectory of a single star, our very own sun. Once again, part of the system is prewired, but part of it requires learning. The prewired bit is a mathematical function that relates the sun's position on the horizon to a bee's orientation—but some of the values of the equation must be set, which is where learning comes in. What the bee learns is a highly specific bit of information about the sun's trajectory at the bee's particular latitude at a particular time of year. A five o'clock winter sun in Boston means something very different from a five o'clock summer sun in California, and a highly focused learning mechanism allows honeybees to take advantage of that information. We know that bees don't simply memorize a correspondence between particular places on the horizon and particular headings, because bees that have been raised in conditions in which they are exposed only to morning light can accurately use the sun as a guide during evening light.

In essence, the bee's azimuth system acts like a sundial run in reverse, and like a sundial, it has to be calibrated. A sundial, which must be oriented with respect to a known compass direction, calculates the time of day based on where the sun is; the navigational centers in the bee's brain calculate where the sun should be based on the time of day. As a consequence, the one thing that bees can't cope with is the discalibration that results from jet lag. In a famous 1960s experiment, Max Renner packed up a hive of bees in Long Island, New York, flew them to Davis, California, and tested their ability to navigate with the sun as a landmark. The jetlagged bees consistently misoriented themselves by 45 degrees, precisely as though they believed it was three hours later.[35] The complex circuitry that allows the bee to use the sun as a guide is built in, but it is not that genes trump the environment (or the other way around). Instead, genes enable creatures to make sensible use of their particular environment. Learning is not the antithesis of innateness but one of its most important products.

Sometimes, even closely related species differ in their aptitudes for learning. The sea slug that I mentioned, *Aplysia californica,* has a rather dim cousin, *Dolabrifera dolabrifera,* that never manages to dishabituate—once it's gotten used to being poked and prodded, it stays that way, even if the prodding has stopped for a long while, whereas *A. californica* will, after a break, play the game all over again. My point is not to cast aspersion on the sticky *D. dolabrifera,* but to make clear that learning doesn't come for free. Whether or not an animal can engage in a particular kind of learning depends on what kind of neural circuitry it has; over the course of evolution, *D. dolabrifera* may have lost one of its learning abilities, the one that tells it that things have finally returned to normal. Learning doesn't come for free; but is a collection of specific mental mechanisms that are the product of evolution.

The variability in the toolkit for learning can be seen even within a group of related species, such as songbird[36]. Some, such as the willow flycatcher, are born with a particular song, and can't learn a thing.[37] For them, rearing makes no difference; they produce the same song with or without a model—they are hardwired (not just prewired) for a particular song.[38] But among songbirds that can learn, there's a world of difference—different learning mechanisms in different species. Others can learn, but what they learn varies widely across species. Some, such as the mockingbird, will pick up on just about anything vaguely songlike that they hear around them—sparrow calls, insect noises, even urban substitutes like car alarms.[39] Parrots will go so far as to imitate human voices.

Still others, such as sparrows and zebra finches, are born with more focused mechanisms that lead them to prefer songs of their own species. Like human babies learning language, such birds seem to break down the "sentences" they hear into the equivalent of phrases and syllables. For a swamp sparrow or a zebra finch, learning a song is a matter of figuring out the local language. For cowbirds, song learning is more a social matter: The males try out a bunch of built-in songs and then stick with the ones that attract the most females.[40]

BORN TO LEARN, HUMAN-STYLE

Given the learning talents of the birds and the bees, it is scarcely surprising that humans, too, come equipped to learn. Like other animals, baby humans are exquisitely sensitive to the statistics of the world that surrounds them. Four-day-old babies can tell the difference between a string of three-syllable words and a string of two-syllable words. In one experiment, Rochester psychologists Jenny Saffran, Dick Aslin, and Elissa Newport presented eight-month-old babies with a long, monotonous string of unbroken syllables like *tibudopabikudaropigo-latupabikutibudogolatudaropidaropitibudopabikugolatu*.[41] Amazingly, infants seemed to extract order from that chaos, using the statistics of syllables to discriminate between "words" like *pabiku* and "part-words" like *pigola*, wherein the only difference was a statistical one: syllables like *pa, bi,* and *ku* invariably appeared as a unit, whereas the syllables *pi, go,* and *la* only sometimes appeared together (at other times *pi* was followed by *da* or by *ti*).

Babies are also able to generalize beyond the information they are given. In my lab, we presented babies with two minutes' worth of "ABA" sentences such as *ga ti ga, ta la ta,* and *ni la ni.* Rather than simply memorizing the sentences they heard, babies seemed to abstract the general pattern, which allowed them to recognize new sentences, such as *wo fe wo,* that followed a different pattern.[42] Neither of these abilities, however, is unique to humans. The ability to detect statistical information is something that all mammals, or even all multicelled organisms, can do, to greater or lesser extents. Less is known about the ability of various animals to generalize more abstract patterns, but the evidence suggests that even in these talents we are not alone: Cotton-top tamarins were able to succeed in exactly the same *wo fe wo* task as the babies in my lab were.[43]

One learning talent that is less common in other animals is the human ability to imitate, with which humans seem to be born. Intriguingly, the brains of animals such as monkeys have a special set of "mirror" neurons that might serve as a prerequisite for imitation.[44] Such neurons fire when, for example, a monkey reaches for a tool, but also fire when the monkey watches someone *else* grasp the tool in the same way. But knowing that someone else is doing the same thing you're doing is not enough to make you imitate them, and it remains controversial whether monkeys can learn through genuine imitation.[45]

It's clear that humans can learn by imitating others, and they start doing so very early in life. Several years ago, University of Washington psychologist Andrew Meltzoff discovered that when he stuck out his tongue at a three-week-old baby, the baby matched him insult for insult, sticking out *her* tongue whenever he stuck out his. Meltzoff later discovered that newborns could do the same, and that they could imitate not just tongue-protrusion (as it is politely called in the academy) but mouth openings and lip pursing.[46]

The yen for imitation may have something to do with something else humans are awfully good at: acquiring culture. Other animals can do it to some extent, but none so richly as humans. A review in *Nature* argued that at least thirty-nine different chimpanzee behaviors are culturally dependent, varying from one group of chimpanzees to the next.[47] Even orangutans seem to have some kind of culture. A just-published study argued that they display at least nineteen different culturally dependent behaviors.[48] For example, among six groups studied, the orangs of Tanjung Putting in Borneo are the only ones to engage in the sport of surfing down falling branches of dead trees, while the Kutai orangs elsewhere in Borneo, along with the Ketambe orangs in Sumatra, are the only ones who seem to have hit upon the love-thyself trick of using tools for autoeroticism. (No word, yet, on whether this is progressively leading them to go blind.)

But animal cultures are not nearly so diverse as human cultures. Although there are differences between animal groups—savanna baboons, for instance, tend to live in large groups, whereas highland baboons tend to live in smaller groups—the overall effects on the lives of individual animals are tiny. Baboons pretty much eat the same sort of diet and have the same sort of daily rituals of child care, food gathering, and looking out for predators wherever they go. Humans, in contrast, differ radically from one place to the next. As ecologist Peter Richerson and anthropologist Robert Boyd have pointed out, in the same range of environments where one can find baboons, there are humans who have radically different kinship systems, social structures, diets, and ways of gathering their daily sustenance, ranging from small-scale hunters who chase small animals with bows and arrows to those who survive mainly by fishing, farming, or raising cattle.[49]

I have occasionally heard psychologists talk as if all it would take to get a baby chimp (or baboon) to act like a human would be a loving human home. But every attempt to raise nonhuman primates in human environments has been a failure; no amount of Head Start will give us a talking chimp or a chimp with one-tenth the cultural variation found in humans in the tiniest corner of Africa. The very ability to acquire culture is, I would suggest, one of the mind's most powerful built-in learning mechanisms.

∞

My own guess is that it is hard to develop a rich culture without a rich communication system. And that brings us to another learning ability no other animal appears to have: the gift for acquiring a communication system with the richness and complexity of language, a system for communicating not just the here and now, but the future, the possible, and the dreamt-of. As I mentioned earlier in this chapter, human newborns *like* speech; they'd rather hear speech than an otherwise similar set of warbles. This may well be an adaptation that assures them of lots of practice, but practice alone is not enough to get a non-human primate to learn language. No matter how much practice you give a monkey, monkey see, but monkey not say.

If monkeys can learn something about the statistics of the world, and even extract abstract regularities, why can't they learn language? It could be that they just don't care to. In fact, dogs (also not known to be gifted language learners) are far more concerned with what humans are thinking than chimpanzees seem to be. Dogs will, for example, watch a human and notice where the human is looking, but chimps don't seem to give a hoot what we might be gazing at.[50] Maybe chimps also just aren't that interested in finding out what we are talking about.

But that's not all there is to the story. One critical difference between us and other mammals is that we are awfully talented at learning new words.[51] Vervet monkeys seem to be born with three different alarm calls: "eagle" (or "look up"), "snake" (or "look down"), and "leopard" ("run into the trees").[52] They get a little better at using those calls over time—a young vervet might make the snake shout when it sees a stick, and older vervets know better—but the vervets do not (at least in the wild) seem to be able to learn new alarm calls. A handful of chimps that have been exposed to sign language have done significantly better, but even for those chimps, learning words seems to be a slow, painful process. Kanzi, the Albert Einstein of chimps (and the only chimp I know to have jammed with both Paul McCartney and Peter Gabriel), produces only about 250 words (lexigrams) after many years of constant contact with her eager caretakers.[53]

The average baby learns that many words before his or her second birthday,[54] and the pace accelerates (perhaps gradually, perhaps quickly, which is a matter of some controversy[55]) as the child learns more. By the time children are in school, they learn something like nine words a day (and without the benefit of formal tuition). Their brains soak up the vocabulary from their environment like a sponge.[56] Human children, unlike chimps of any age, are able to use what they know about one word (or set of words) to help them with another. Psychologist Ellen Markman has shown that if a researcher says "dax" while a two-year-old looks at a spoon and a garlic press, the two-year-old (who presumably already knows the word "spoon") will guess that "dax" refers to the garlic press.[57] In a 1957 experiment that helped launch the modern study of language acquisition, the late Roger Brown showed that children know that if you say, "Can you see a *sib*?" you probably have in mind an object, whereas if you say "Can you see any *sibbing*?" you probably have in mind an action or a process.[58] No other mammal seems to be equipped to use such clues for word learning.

Even more dramatically, no other species seems to be able to make much of word order. The difference between the sentence "Dog bites man" and the sentence "Man bites dog" is largely lost on our nonhuman cousins.[59] There is a bit of evidence that Kanzi can pay attention to word order to some tiny extent, but certainly not in anything like as rich a fashion as a three-year-old human child.[60]

The mechanisms that allow children to learn language are so powerful and resilient that children can acquire the basics of grammar in almost any circumstance. Middle-class Western parents may dote on their children's every word, and sometimes even give their kids explicit language lessons, but parents don't have to do that to get their children to talk. In some cultures, parents think it's silly to talk to young children, and their children still manage to learn language just fine.[61] Kids learn more words if they hear more words,[62] but all normal children, regardless of circumstance, learn the basics of their language in a way that chimps could only envy.

Why are humans specially able to draw connections between meanings and strings of words? Within the scientific community, this question remains controversial. Chomsky, Pinker, and others have argued that we have an innate universal grammar (or "language acquisition device") that tells a child what kinds of languages are possible;[63] others have argued that humans have little more than an inborn facility for rapid auditory processing. In my view, the theory that all that is special about language is our facility for fast auditory processing is highly implausible when one considers the extent to which humans vastly outstrip chimpanzees even when it comes to sign language. If there is a language acquisition device, and I suspect there is, it is likely to depend on a mix of capacities, some specialized for language, others (such as the ability to learn abstract rules and the ability to collect statistics) used more widely throughout mental life.

The debate is not really about whether language is innate ("built in") or learned; it is really about the extent to which the mechanisms that allow us to learn language—which themselves presumably are innate—are specialized for language. Whatever the "language instinct" consists of, it is not a particular language (no child is born knowing English, Hindi, or Japanese) but a particular *built-in* way of acquiring new information. Language is perhaps the most powerful example of what you can do if you are born with the right kinds of mental machinery for learning.

3

BRAIN STORMS

Brain: an apparatus with which we think we think.
—Ambrose Bierce

As THE BASEBALL writer Bill James once observed, "If an elephant walks through the snow, it leaves footprints." If our minds are well organized at birth, ready to understand the world and ready to learn, we ought to be able to see reflections of that structure in the brain of a newborn. And indeed we can.

To a first approximation, a baby's brain looks like a miniature version of the brain of an adult. It has the same organization into hemispheres and lobes, the same peaks and valleys known as *gyri* and *sulci,* the same divisions of the all-important cortex into a six-layered sheet, and the same basic pathways from the senses to the brain. If there is a lot of work still to be done, it is also clear that by the time a baby leaves the womb, the overall structure of the brain is already in place. (Comedian Steven Wright claims that he kept a diary shortly after birth. "Day One. Still tired from the move. Day Two. Everybody talks to me like I'm an idiot.")

The fact that the brain is well organized at birth is not by itself a knockdown argument for a prewired mind. Even if some neural structure is present at birth, it could in principle still be the product of learning. Babies can learn even when still inside the womb. In one

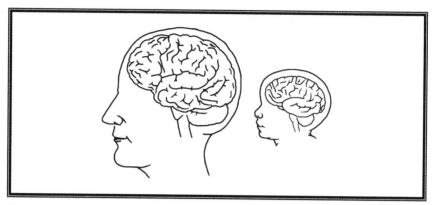

Figure 3.1   Brains of an adult and a newborn
Illustration by Tim Fedak

delightful study, psychologists Anthony DeCasper and Melanie Spence asked prospective mothers to read aloud a three-minute passage from either Dr. Seuss's *The Cat in the Hat* or Nancy and Eric Gurney's *The King, the Mice, and the Cheese* during the last trimester of pregnancy. Tested just a day or two after birth, the infants who had been exposed to Seuss in the womb preferred Seuss; those who heard *The King* preferred *The King*—even when another person read the stories.[1] This is not to say that late-trimester infants actually *understood* the Cat's tale, but they did seem to have caught on to its distinctive rhythms. Another study showed that third-trimester fetuses could pick up the melody in *Mary Had a Little Lamb,* and another that they could recognize the melody from the theme song of a British soap opera.[2] (I am not, however, suggesting that you try this at home. There's no evidence that prenatal exposure has any lasting long-term consequences, and some experts believe that such deliberate exposure could actually be disruptive to the developing auditory system as well as to the baby's natural wake-sleep cycles.[3])

Showing that babies can learn in the womb does not resolve the prewiring question either, because there could still be prewiring before rewiring, and learning the rhythms to *The Cat in the Hat* is not the same thing as dividing the brain into hemispheres or wiring the connections between two cortical areas. The strongest test of whether a certain aspect of neural organization is built in—in the sense of not depending on experience—is to take that experience away. Scientists have been devising subtler and subtler ways to do that for decades, always with pretty much the same results: The initial organization of the brain does not rely that much on experience; the nativists are right to think that the slate is not blank. But you wouldn't know it if you read most psychology textbooks. When the subject of experience and early brain development comes up, the textbooks almost always point to the same experiment—and they almost always tell only half the story.

The story they tell starts with pioneering experiments conducted in the 1960s by two Harvard neurophysiologists, David Hubel and Torsten Wiesel.[4] In normal kittens (and in most primates), information from the left and right eyes converges (after a trip through the midbrain) onto a layer of the visual cortex in a pattern of alternating "columns" or slabs, stripes of neurons known in the trade as "ocular dominance columns." About half a millimeter wide, these stripes switch back and forth, a slab for the left eye, a slab for the right eye, illustrated here as dark bands for the left eye, light bands for the right.

Hubel and Wiesel used an eye patch to investigate what would happen if kittens were deprived of visual experience in one eye. What they found was that for the kittens with the eye patches, the neat pattern of alternating neural stripes disappeared. The amount of brain tissue devoted to the open eye expanded, and the amount devoted to the covered eye shrank. Textbooks often stop there and use this result to argue that early brain structure is the product of experience.

What the texts usually don't describe is what happened to a set of kittens that were deprived of visual experience in *both* eyes from birth. In those kittens, the striped ocular dominance columns formed pretty much normally. Taken together, what these two studies show is not that ocular dominance columns are the pure product of experience, but that they develop *in two stages*: a period of initial organization that does not require experience, and a later stage of fine-tuning that does—rough draft followed by calibration.

As Hubel later put it, "The nature-nurture question is whether postnatal development depends on experience or goes on even after birth according to a built-in program. . . . The unresponsiveness of

Figure 3.2   Ocular dominance columns
Illustration courtesy of Estela O'Brien and Ehud Kaplan, Mt. Sinai School of Medicine

cortical cells after deprivation was mainly due to a deterioration of connections that had been present at birth, not a failure to form because of a lack of experience."[5] In other words, nature is powerful enough to shape neural architecture in advance of experience, but also supple enough to adapt that architecture to unusual conditions—nature once again providing a way of making sensible use of nurture.

A whole raft of more recent studies points in the same direction. Nature provides a first draft, which experience then revises. Neuroscientists Jonathan Horton and David Hocking showed that in primates, ocular dominance columns actually form in the dark of the womb. German researchers Imke Gödecke and Tobias Bonhoeffer raised kittens in such a way that both eyes would have experience—but not at the same time. When one eye could see, the other was sutured shut, and vice versa. If experience were doing all of the work in fine-tuning the visual cortex, one might expect that the organization of the "orientation maps" of the two eyes (brain circuits that process the orientations of lines) would be different, reflecting the likely differences in experience between the two eyes. But Gödecke and Bonhoeffer found that the organization of the two brain maps was essentially identical.[6]

Another great rallying cry against innateness has been the three-eyed frog. There's no such beast in nature, but in the late 1970s,

Martha Constantine-Paton created them in the laboratory, by transplanting the embryonic eye primordium of one frog into a normal, two-eyed frog. The amazing result was that the three-eyed frogs developed ocular dominance columns, something that ordinary frogs do not even have (presumably because the alternating stripes are a brain's way of dealing with overlap between two forward-facing eyes; ordinary frogs do not need to deal with this overlap because their eyes face opposite directions).[7]

Many people seemed to assume that the ocular dominance columns of the three-eyed frogs were wired up on the basis of visual experience, but a more recent study by two Australian neuroscientists, Sarah Dunlop and Lyn Beazley, showed that one can get the same effect in a marsupial mouse, the fat-tailed dunnart *(Sminthopsis crassicaudata).*[8] What's interesting is that the dunnart, among the most immature mammals at birth, did all this neural reorganization before it could see, while its eyes were still buried beneath its skull. *Internally*-generated neural activity (something I will discuss in Chapter 6) plays an important role here,[9] but no visual experience is required. Rather than providing evidence for the importance of the external environment, the three-eyed animals illustrate an ancient, built-in trick for wiring up competing inputs.

Still more recently, Duke neuroscientists Larry Katz and Justin Crowley surgically removed the eyes of newborn ferrets, thereby cutting the ferrets off from any possibility of visual experience. Several months later, they tested to see whether the ocular dominance stripes had formed normally. Consistent with Hubel's interpretation that the initial formation of the columns is independent of experience, removing the retinas made no difference: The striped columns developed just fine, even when the ferrets couldn't see a thing.[10]

Visual information normally follows a pathway in the brain from the retina through a structure known as the thalamus and on to the cortex. Another way of testing the importance of experience in the early organization of the cortex is to interrupt this pathway.[11] John Rubenstein's lab at the University of California at San Francisco did just that, creating mice with a genetic mutation that prevented neurons in the thalamus from reaching their usual destinations in the cortex. If signals from the thalamus were crucial to cortical organization, the mutation should have led to markedly abnormal cortical development. Instead, at birth, the mutants with disrupted thalami appeared to have *normal* cortices, indistinguishable on most measures from those of normal mice! (About a fifth of the mice did show abnormalities, but those abnormalities—things like rips on the boundary between cortex and subcortex—probably occurred for structural reasons, not because of the lack of thalamic input per se.) What this study tells us is that an awful lot of early brain organization can proceed normally, even without the usual experiential inputs.

A group of neurophysiologists led by Thomas Südhof went even further, finding a way to shut off most learning altogether, not just from the visual world, but from all the senses.[12] Most learning is thought to depend on electrical communication across "synapses" that join neurons. In the course investigating the molecular basis of that communication, Südhof's team discovered a way to silence that synaptic communication by genetically engineering mice that lacked a protein vital for neurotransmission. Although the team was expecting to find radical differences between the normal mice and the mutant mice, up until birth, the mutation didn't seem to matter (after that it was deadly, since without synaptic transmission there is no breathing, let alone learning). As in the study of thalamus disruption, no matter what the researchers looked at—from the segregation of layers of brain matter to the properties of the synapses that connect neurons—they couldn't find any differences between the brains of normal mice and the brains of the mutants. As far as they could tell, the brain does a pretty good job of assembling its initial structure even when a congenital mutation renders most forms of learning impossible. Two further studies have shown that these results do not hinge on the particular way in which the Südhof group silenced synaptic transmission. There are other ways of silencing synapses, but these, too, cause no obvious alteration in initial neural organization.[13] The basic structure of the young brain depends only minimally on experience.

BRAIN PUTTY

If these were the only studies of a young brain, we might wonder what all the fuss was about. Why would anyone doubt the nativist position at all? In the studies of early development that I have just described, experience seemed to have hardly an impact at all. But there is another, equally impressive set of experiments that seems—at first glance—to point to an almost opposite conclusion: that certain kinds of experience can radically alter brain organization. This second group of studies suggests that the fabric of the brain may be like putty.

One of the first such studies was done in the 1980s by Dennis O'Leary of the University of California at San Diego, who literally transplanted neurons from one part of the brain to another. Working with newborn rats, O'Leary's team took neurons from the part of the brain that interprets touch (the *somatosensory cortex*) and stuck them into the part of the brain that handles vision (the straightforwardly named *visual cortex*).[14]

It is perhaps not surprising that the transplanted cells survived the move, but what was remarkable was that the somatosensory cells that were transplanted into the visual cortex took on an altogether new identity, behaving in many ways as if they had always been visual neurons instead of somatosensory neurons. For example, they grew connections to the superior colliculus, a subcortical way station for visual information, rather than to the spinal cord. (Actually, although I've reported this experiment the way that it is generally described, there's an important oversimplification. O'Leary's experiments weren't really conducted with full-fledged somatosensory neurons but with what developmental biologists call "presumptive" somatosensory cells, primordial cells that would under ordinary circumstances become somatosensory cells.)

This was no one-way transformation, either. O'Leary and his colleagues found that they could also run the transplants in reverse, taking presumptive visual neurons and sticking them into the somatosensory cortex. *Those* transplanted cells behaved like somatosensory cells, growing connections into the spinal cord, showing once again how malleable primordial neurons can be. A later set of experiments, by neuroscientists Ole Isacson and Terrence Deacon, showed that brain cells can sometimes even be transplanted from one *species* to another. In cell transplants from fetal pigs to adult rats, for example, the transplanted neurons often grew connections as if they were normal rat neurons.[15] Clearly, a neuron's fate is not fixed from the moment it comes into being.

Other scientists have gone even further, asking whether young brains can still adapt and continue to function after large chunks are removed. For example, in adult monkeys, an important part of the job of recognizing objects takes place in a part of the inferior temporal cortex known as area TE. Without it, adult monkeys lose much (though not quite all) of their ability to recognize objects. In contrast, the brains of young monkeys are remarkably resilient. In one study, a team of scientists removed the TE area from newborn monkeys. Within ten months, their brains had recovered sufficiently such that the TE-less monkeys were able to recognize objects nearly as well as normal monkeys, apparently by shifting the burden of object recognition to parts of the brain (mostly nearby) that were uninjured, a truly remarkable demonstration of the amazing flexibility of the growing nervous system.[16]

In an even more radical bit of brain surgery, MIT neuroscientist Mriganka Sur and his students managed to create ferrets in which visual input was rerouted to the auditory cortex. By surgically removing the superior colliculus (the ordinary target of visual input from the retina), as well as some of the ordinary inputs to the medial geniculate nucleus (the MGN, a part of the thalamus that usually receives auditory input), the Sur lab was able, in the memorable words of a grad school classmate of mine, to connect "the eye bone" to "the ear bone."[17]

In fact, there are lots of ways to get the "ear bone" to respond to visual input. Sur showed that the auditory cortex responds to visual input in rewired ferrets. Essentially the same thing happens in animals that have been deafened at birth.[18] The reverse holds true as well: The "visual" part of the brain responds to auditory input in kittens that have been raised with sutured-shut eyelids.[19]

What's true of the animals seems to be true of people, too. No scientist would deliberately remove part of a human's brain (or rewire it) just to satisfy scientific curiosity. But brain scans of humans born with sensory impairments closely parallel the results of the animal studies. For instance, Oregon psychologist Helen Neville has shown that in human adults born deaf, parts of the brain that are ordinarily devoted to hearing respond to visual stimuli.[20] Other researchers have found that the "visual cortices" of babies who are born blind respond to auditory input.[21] It has even been shown that the visual cortex becomes active when blind people read Braille.[22]

More evidence of the resilience of the young human brain comes from studies of children who have suffered brain injuries. Infants who suffer severe brain damage at birth have often been known to develop near normal cognitive functioning.[23] A few young children have had to have their entire left hemispheres removed (a rare, radical surgery that appears to be the only way to prevent certain life-threatening seizures). Astonishingly, these children learn to talk more or less normally, shifting language function to the right hemisphere.[24] In short, young human brains, like those of our animal cousins, are often (though by no means always) remarkably able to reconfigure themselves, on-line.

Adult brains aren't as "plastic" as the brains of infants, but even here there is some room for change. University of California at San Francisco neuroscientist Michael Merzenich discovered that adult monkeys can reallocate parts of their cortex when those parts are no longer needed for their original function. In a study in which Merzenich amputated the third finger of a monkey (a gruesome experiment that could have huge payoffs in treating strokes and spinal cord injuries), he found that the parts of the cortex that were originally devoted to the amputated finger gradually (over a period of months) began to respond instead to the neighboring (intact) fingers.[25]

HEAL THYSELF

Rewiring, rerouting, reconfiguration. What do all these examples mean for our questions about how the brain and mind develop? To scholars such as University of California at San Diego cognitive scientists Elizabeth Bates and Jeffrey Elman, they are the death knell for "nativist" theories holding that children are born with significant mental structure. Bates, for example, has argued that the "plasticity" findings have "led most developmental neurobiologists to conclude that cortical differentiation and functional specialization are largely the product of input to the cortex . . . provid[ing] a serious challenge to the old notion that the brain is organized into largely predetermined, domain-specific faculties,"[26] and neuroscientists Steven Quartz and Terrence Sejnowski have used them to argue that "nativist theories appear implausible."[27]

But, logically speaking, there's no reason to see plasticity as being in conflict with the idea of built-in structure. "Built-in" doesn't mean unmalleable; it means organized in advance of experience. What plasticity tells us is not that embryos need experience to form the initial structure of the brain—but rather that the initial structure can be *changed* afterward in response to experience. The two—initial formation and later revision—are, of course, logically independent; whether a system can change itself is separate from the question of where it gets its initial structure.

The terms "built-in" and "unmalleable" often get confused, perhaps because minds, like computers, process information, and early computers relied heavily on built-in circuitry that was "hardwired" and unchangeable. But there is no necessary equation between the two. Engineers have long since moved on to reprogrammable "firmware" that is programmed at the factory but always changeable, updateable with the latest version from the web. Evolution may have caught on far earlier: Just because something is preprogrammed doesn't mean it can't also be reprogrammed. In many systems, the brain may well use a mix of internally generated cues to prewire and environmentally generated cues to rewire.

When I say that the brain can recover from an injury, it sounds exciting; if I say that the body can recover from injury, however, no one is likely to be impressed. Even little children know that skinned knees are rarely fatal. We can also recover from broken bones, bruises, boils, burns, welts, pimples, and even broken hearts. We're not immortal, not invincible—high-speed automobile crashes can do us in, but the human body has dozens of tools for self-repair.

The ability to regrow language after early brain damage is just one of them. Taken in this broad perspective, the fact that the brain can recover from injury is hardly surprising. In fact, perhaps the only real surprise here is how *inflexible* the brain is. Most parts of the body constantly replace their cells, whereas the adult brain's stock of neurons is almost entirely fixed.[28] Your liver cells are constantly replenished, but your brain must largely (though not entirely) make do with the neurons you had when you were born,[29] effecting its repairs mainly by modifying the wiring between neurons rather than by generating new neurons. Still, the brain can, like most other parts of the body, manage a large degree of self-repair. (One form of self-repair takes advantage of built-in redundancy—if one kidney is lost, the body can shift function to the other; if one hemisphere is lost, at least some function is transferred over to preexisting counterparts on the opposite hemisphere.[30])

Recovery from injury is something that happens during the everyday life of most individuals; transplant studies are not. Outside of the lab, cells don't get transplanted from the somatosensory cortex to the visual cortex. Yet the body seems just as adept at dealing with this challenge. Here again, there's nothing special about the brain. Take a cell that would ordinarily become an eye cell, stick it in the stomach, and it will turn into a stomach cell.[31] That doesn't mean that the presumptive eye cell has *learned* how to be a stomach cell.

Instead, as we will see shortly, this kind of plasticity is a consequence of the very procedures by which the body assembles itself. Most cells in the body (except mature red blood cells and platelets) are born with a *complete* set of instructions—instructions about how to behave should it be called upon to be a stomach cell, instructions about how to behave should it be called upon to be an eye cell, and so forth. *Which instructions a given cell follows is partly determined by its neighbors.* Being surrounded by stomach cells can lead an impressionable young cell to act like a stomach cell. The same thing, I would suggest, is happening in the brain cell transplants. A visual-neuron-to-be that hasn't yet gone full steam into following the built-in instructions for visual neurons may be able to change gears and start following the plan for somatosensory cells. Adult neural stem cells can even turn into blood cells, clearly not something that should be attributed to learning.[32] This isn't something special about brain development; it's simply how developmental rules—genes—work.

But why did the ferret's retina cross the auditory road? Although people often describe Sur's famous experiments as experiments in rewiring, Sur did not literally rewire the ferrets, which is to say that he did not literally connect the output "wires" of retinal cells (axons) to the neurons in the auditory thalamus. Instead, Sur goosed things a bit and then let the axons find their own way. Understanding how the experiment actually works helps to understand what's really going on.

Here's how it worked. Retinal axons usually try to wire themselves to the superior colliculus. Sur's team kept them from doing that by removing the superior colliculus. If that were all that Sur and his colleagues had done, though, the rewiring might not have worked, for the retinal axons might have been left with nowhere to go. To complete the rewiring process, Sur removed the inputs that ordinarily feed into the auditory thalamus.[33] This last step—which further experiments have shown to be most crucial[34]—seems to have caused the auditory thalamus to send out a signal saying, in essence, "connect to me." Since the retinal axons were at that very moment looking for a partner, it was a match made in laboratory heaven.

As we will see in Chapter 6, the sorts of signaling systems that might bring together retina and auditory thalamus are widespread throughout the brain. It might be that sensory input plays a role in the rewiring experiments—that's what the antinativists want you to believe—but it might not. Instead, the rewiring might have happened entirely on the basis of the interactions between the feelers of the wandering retinal axons and the signal molecules sent out by the expectant auditory neurons. I would not be at all surprised if one could get the same rewiring in a ferret raised in the dark.

BETWEEN NATURE AND NURTURE

So where does this leave us? We still don't know exactly how "plastic" the brain is. Like many eager scientists, the latter-day antinativists have overstated their case. They write not only of flexibility (clearly true), but of an "equipotentiality" in which any part of the cortex could grow into any other (wild exaggeration). For example, in what seems like a compelling passage, San Diego cognitive scientist Elizabeth Bates wrote, "Isacson and Deacon (1996) have transplanted plugs of cortex from the fetal pig into the brain of the adult rat. These 'foreigners' (called 'xeno-transplants') develop appropriate connections, including functioning axonal links down the spinal column that stop in appropriate places. Although we know very little about the mental life of the resulting rat, no signs of pig-appropriate behaviors have been observed."[35]

But Bates did not report even more striking experiments by biologist Evan Balaban, who implanted parts of quail midbrains into chick embryos, creating animals known as "chimeras" (in honor of the lion-headed, goat-bodied, serpent-tailed beast of Greek myth). The chimeras walked like chickens but crowed like quail, showing that not every cell takes on the properties of its new home.[36] As Balaban's studies and other recent work has shown, not every transplant takes on properties of its new surroundings. Some, especially those transplanted later in development, reflect their source rather than their new destination, especially when such experiments are performed later in development.[37] Moreover, the transplant studies may rely heavily on the similarities between donor and host.[38] Although experiments do show that one can sometimes move sensory tissue from one sensory area to another, they have never shown that one could, say, take a sensory cell and have it integrate properly into, say, a motor area, the amygdala (the seat of emotion[39]), or the prefrontal cortex (which appears to play a central role in our ability to make decisions[40]).

Similarly, as impressive as Sur's ferret studies are, they are not without problems. The "rewired" ferrets could not see as well as normal ferrets and showed a peculiar preference for horizontal lines that normal ferrets do not show. Moreover, a fifth of the neurons in the auditory cortex did not respond to visual input, and the ferrets had brain maps that were disorganized compared to those of normal ferrets.[41] Although experiments have shown that one can get visual input to feed into the auditory cortex, they have not shown that visual input can be rerouted anywhere in the brain willy-nilly. The early stages of visual and auditory processing are quite similar,[42] and the relative success of rewiring may well have depended on that similarity.[43] (Steven Pinker has suggested that plasticity may go no further than this, noting that there have been no demonstrations of exchanges between nonsensory areas; John Kaas, a leading neuroanatomist at Vanderbilt, has said that he expects to find even more plasticity at higher cortical levels that direct more complex behaviors, but he also argued that we simply do not yet know enough about the cortex to be able to tell.[44])

Things are similarly murky when it comes to recovery from brain damage. Recovery from brain injury is indeed much more dramatic in infants than in adults, but even in infants recovery may be only partial. Although it is true that children whose brains are damaged early in life often recover to a remarkable extent, it is also true that such children face lasting deficits. For example, neuroscientist Faraneh Vargha-Khadem of London's Institute of Child Health described a case study of a girl who suffered bilateral damage to her hippocampus at birth. In adults, damage to the hippocampus causes problems for long-term memory. In the famous case of the patient known as HM, removal of the hippocampus (among other adjacent areas) left him completely unable to store new facts in long-term memory.[45] The child with similar damage was able to learn and remember quite a bit, including the meanings of words, but her spatial abilities, her temporal abilities, and her memories of specific events were all profoundly impaired.[46] Disorders such as cerebral palsy (which affects posture and movement),[47] autism (which affects social cognition and communication),[48] and dyslexia (a reading disability)[49] are (at least for now, given limits on contemporary medicine) lifelong propositions. Especially where disorders stem from genes rather than from trauma, my hunch is that there may be significant limits on plasticity.

No matter what the final answer looks like, it is already clear that the nativists and the antinativists are both right about something. The nativists are right that significant parts of the brain are organized even without experience, and their opponents are right to emphasize that the structure of the brain is exquisitely sensitive to experience. Nature has been very clever indeed, endowing us with machinery not only so fantastic that it can organize itself but also so supple that it can refine and retune itself every day of our lives. What we will see in the remainder of the book is that both of these properties are direct and natural consequences of the elegant biological processes that direct the development and maintenance of the brain.

4

ARISTOTLE'S IMPETUS

In all things of nature there is something of the marvelous.

—Aristotle

MANY PEOPLE THINK of genomes—the collections of genes that make each individual and each species unique—as blueprints, little DNA maps for the growing organism. It's not a bad metaphor as far as it goes: There is some sense in which DNA provides a plan for a growing embryo. But as we've seen, in many respects the blueprint metaphor is way off. There's nothing in your genome that corresponds to a picture, no simple connection between the parts of a genome and the parts of a brain. In some science-fiction world in which genes really were blueprints, science and medicine would be more straightforward. Doctors would be able to look at a disruption in the genetic code and immediately see how it would affect the brain, or look at a congenitally disordered brain and instantly know where in the genome to look for the problem. But on our planet, genomes don't work that way. My goal in this chapter is to explain their true function.

I begin, curiously enough, with Aristotle (who didn't know from genes). Aristotle's contribution, derived from his own dissection of chick embryos, was his realization that embryos emerge gradually in a series of stages via a process we might think of as "successive approximation." (Biologists sometimes call this process *epigenesis, epi* meaning "on top of" and *genesis* meaning "generation," but the term has come to mean so many different things that I will not be using it here.)

In his essay "On the Generation of Animals," Aristotle suggested that, rather than being fully formed from the beginning, the body is built incrementally: "The upper half of the body, then, is first marked out in the order of development; as time goes on the lower also reaches its full size. . . . All the parts are first marked out in their outlines and acquire later on their color and softness or hardness, exactly as if Nature were a painter producing a work of art, for painters, too, first sketch in the animal with lines and only after that put in the colors."

Although not every detail of Aristotle's theory was right (he believed, for example, that bones were first made of what he called "seminal residue"), his basic premise was correct: Every embryo goes through an orderly progression of stages in which organs get laid down first in a very rough form and then are refined in successive waves of more and more detail.[1] In contrast to the misguided claims of the seventeenth-century preformationists (who, recall, thought the embryo was fully formed), real embryos do not look a thing like the adults they are destined to become. A two-day-old embryo certainly looks like nothing so much as a mulberry, and the human embryo four weeks after conception arguably looks more like a fish than a person. (The rumor that "ontogeny recapitulates phylogeny" is, however, a lie. The famous phrase, coined by German biologist Ernst Haeckel, suggested that embryos went through stages of development that traced out the history of our evolutionary development. That slow-to-die fallacy[2] may have reached its apex when famed childcare expert Dr. Spock once claimed, referring to embryonic pharyngeal arches that look like gills, that "each child as he develops retrace[s] the whole history of mankind, physically and spiritually."[3] But human embryos don't really have gills. To the extent that early human embryos resemble other organisms, what they resemble is not the *adult* form of those organisms but the *embryos* of those organisms, as properly surmised by Haeckel's less quotable predecessor Karl Ernst von Baer.[4])

The bottom line is that we now realize that embryos unfold in a series of stages. During conception, sperm and egg unite to form the

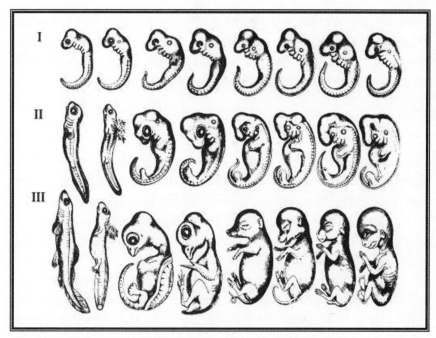

Figure 4.1   Stages of embryonic development in eight species
Drawing from Ernst Haeckel (1866)

fertilized egg known as a *zygote*. The zygote, which begins as a single cell, soon divides into two cells, and then each of those cells divides again, and again, ultimately forming a ball of eight nearly identical cells. The ball of cells eventually flattens, layers of cells start to form, and soon cells start to take on specialized fates, limbs begin to bud, and organs begin to blossom. And, as we will see in the next chapter, the principle of successive approximation (or gradual unfolding) applies as much to the development of the brain as it does to the rest of the body.

For all that Aristotle did surmise—stunning, given what he had to work with—his theory was still hopelessly incomplete. Aristotle could still not address some of the most important questions. What is it that drives embryos from one stage to the next, and what is it that drives an embryo to follow in its parents' footsteps? Why does a monkey embryo grow up to be a monkey rather than a grapefruit? Aristotle couldn't answer, except to suggest that there was some "motive force."

Aristotle's explanation was like saying that cars move forward when the axles rotate—without mentioning a thing about gasoline or the crankshaft that drives the gearbox that turns the axles. And that's where genes—the true motive force—come in.

WHAT'S IN A GENE?

The modern scientific conception of what genes do—and how they supervise the construction of the body and brain—began more than 150 years ago with a view that I will call the Trait Theory. Scientists later elaborated this idea into ever more complex theories that I will call the Enzyme Theory, the Protein Template Theory, and finally, the (most complete) Autonomous Agent Theory.

In the paragraphs to come, I will explain each one, giving a conceptual history of the gene. This excursion into the history of modern biology may at first seem like a diversion. Why should we care how scientists discovered what genes do, if what we really care about is the development of the mind and brain? Why should it matter to us whether genes are for traits or for enzymes? And what difference does it make whether genes act as templates or as autonomous agents?

We should care because answers to questions about the interrelationship between nature and nurture, and why it is difficult to separate the two, can only be truly understood with reference to the way in which biological structure is actually built. Without understanding the true nature of genes, we cannot hope to understand what makes it challenging to diagnose and treat disorders, how the genome gets so much out of so few genes, or how it evolves new cognitive systems from old.

Let us begin with the Trait Theory, one of the simplest theories one can imagine. The idea is that each gene affects a single trait—say, height, eye color, or IQ. Although, no current biologist takes this explanation as a complete and accurate account of what a gene is, the one gene, one trait theory deserves mention because vestiges of it appear all the time—in newspaper articles ("Migraine Gene Discovered"), books (*The Hungry Gene*[5]), and professional articles in fields outside of molecular biology.

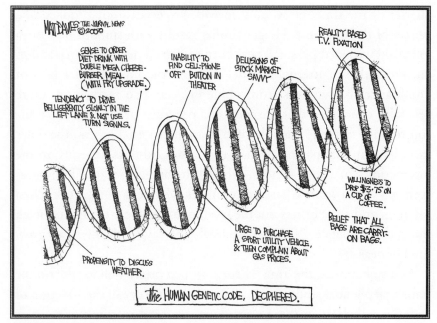

Figure 4.2   The Trait Theory, updated for the twenty-first century
Reprinted by permission of Matt Davies and Tribune Media Services

The Trait Theory, which was pivotal in the subsequent discovery
that DNA represents the molecular basis of heredity, and everything
gene-related that has followed, began in the 1860s with the Austrian
monk Gregor Mendel and his painstaking studies of the mechanics
of inheritance, of what makes children resemble their parents. Focusing on the common garden pea *(Pisum sativum)*, Mendel bred more
than 28,000 plants—tall ones, short ones, smooth ones, wrinkled
ones, green ones, yellow ones—about seventy different purebred varieties in all.[6]

In the course of his studies, Mendel quickly refuted the then popular idea that offspring were simple blends of their parents. Crossing
purebred yellow peas with purebred green peas, for example, invariably led to yellow peas rather than to yellowish-green blends. When
Mendel bred together two hybrid yellow peas (that is, two yellow
peas, each born of a cross between a purebred yellow pea and a purebred green pea), something even more astonishing happened: About a

quarter of the time, he wound up with *green* peas—that looked like *neither* of their parents. He got similar results with smooth peas born to hybrid wrinkled peas, tall peas born to hybrid dwarf parents, and so on. From these patterns, Mendel deduced the now familiar laws of dominance (brown eyes trump blue eyes, and so forth) and, more generally, concluded that the color of a pea must depend on two trait-controlling "factors," one inherited from its paternal parent, the other from its maternal parent. Which trait a pea exhibits depends on the interactions of those factors, which ultimately became known as "genes."[7] (In a John Chase cartoon, it's Mendel's night on kitchen rotation, and as he brings out the food, one monk can't help but break the code of silence. "Brother Mendel," he says forlornly, "we grow tired of peas.")

As it so happens, the Trait Theory worked pretty well for the simple traits that Mendel studied, but as a complete theory of the function of genes, it fell short in three ways. First, we tend to think of traits as qualities that vary from one individual to another: I have brown eyes; you have green eyes. But most genes have nothing to do with differences between people; the vast majority of them are shared by all normal individuals. Genes do a lot more than just shape differences among individuals, and the trait theory does not explain why. Second, most traits are influenced by more than one gene; skin color, for example, is influenced by at least thirty.[8] Finally, it is not uncommon for a single gene to influence several different properties, sometimes not obviously related, as in the single gene that leads to two of the most distinctive features of Siamese cats—their unusual coloration (light body, dark extremities) and their crossed eyes.[9]

Although Mendel's ideas were largely ignored for thirty-five years,[10] they eventually led to the Enzyme Theory, which was a big leap because it was the first mechanistic theory of how physical entities could influence heredity. In 1902, a few years before the term "gene" was coined, a British doctor named Simon Garrod discovered that certain disorders—which he later dubbed "inborn errors of metabolism"[11]—ran in human families according to patterns that closely resembled

Mendel's pedigrees of inheritance. Whether a child inherited disorders such as albinism (which makes skin, hair, and eyes white) or alkaptonuria (a frightening but harmless disorder that turns urine black) could be predicted with great accuracy using Mendel's factors.

These observations gave rise to the idea that genes exerted their effect by *influencing the production of enzymes.*[12] Enzymes are biological catalysts that make chemical reactions go faster, and the Enzyme Theory suggested that physical (and presumably mental) disorders were the product of particular missing enzymes. The urine in patients with alkaptonuria turned black because they lacked an enzyme known as homogentisic oxidase; the skin of the albino patients was white because they lacked the enzyme that catalyzes the process of turning tyrosine into skin pigment. Further evidence favoring the Enzyme Theory came in the 1940s, when two Caltech biologists, Edward Beadle and George Tatum, used microwave irradiation to systematically generate strains of mutant bacteria, each of which, again, appeared to lack particular enzymes.

According to the Enzyme Theory, most famously captured in Beadle and Tatum's memorable slogan "one gene, one enzyme," *all* congenital disorders (mental or physical) were understood as being the products of missing enzymes. And some really are: Phenylketonuria (PKU), for example, is a form of mental retardation that results from a missing enzyme, and if caught early enough it is relatively amenable to treatment, as a matter of managing diet and dietary supplements. But the Enzyme Theory actually missed much of the larger truth: A great many genes (and corresponding disorders) have nothing to do with enzymes and simply do not fit into the enzyme framework. Disorders from sickle-cell anemia to Lou Gehrig's disease (amyotrophic lateral sclerosis) to muscular dystrophy and cystic fibrosis have nothing to do with enzymes, yet everything to do with genes.

In the 1950s and 1960s, scientists discovered that genes do much more than simply guide the construction of enzymes. Most enzymes belong to a larger group of molecules known as proteins, and it turns out that genes are pivotal in the construction all of the members of that much larger group of molecules.

Proteins are long chains of twenty or so basic molecules known as amino acids that are twisted and folded into complex three-dimensional structures such as fibers, tubes, globules, and sheets. Amino acids, in turn, are particular arrangements of carbon, hydrogen, oxygen, and nitrogen atoms. (Your body makes many of these amino acids, but nine are "essential" because they can only come from your diet. Animal meats typically contain all the missing amino acids, but many plant products do not—lysine, for example, is absent in many grains—which is why vegetarians must carefully balance, or "complement," their sources of protein.)

There are literally hundreds of thousands of different proteins in a human body.[13] An average cell has thousands different proteins, and, all told, they make up more than half the body's dry weight.[14] In addition to enzymes, there is a huge range of other proteins. For example, keratin (the principal protein in hair) and collagen (the principal protein in skin) help to build the structures of the body. Others, such as prolactin and insulin, are hormones used for communication between (and within) organs. Still others serve as everything from motors to couriers (such as hemoglobin, which exists to transport oxygen). And then there are *channels,* complex cellular gates that open and close to control the flow of molecules into and out of a cell, and *receptors,* receivers of biochemical signals that can be thought of as sentinels that capture messages and convey their content to the inside of a cell without letting the messengers themselves through the membrane walls. Proteins are involved in just about every aspect of life.[15]

The first step toward the Protein Template conception of genes, which held that genes were involved in all proteins, not just enzymes, came in the 1940s. Up until then, many scientists thought that genes were just one more special kind of protein, but in 1944 a largely unsung American biologist named Oswald Avery discovered otherwise. His great advance came in a study of the uncomfortably familiar bacterium we know as *pneumococcus.* Pneumococcus comes in two varieties, a lethal, "smooth-coated" S strain and a normally harmless, "rough" R strain, so named for their appearance under a microscope. In the late 1920s, British biologist Frederick Griffiths discovered that heat-killed S strain (which on its own was not lethal) could "transform" normally safe R-strain bacteria into deadly killers. But Griffiths was not able to explain why. Avery cracked the case by a process of elimination, ruling out, one by one, all the substances contained within the S strain until the only substance left was a mysterious sticky acid that had first been identified in 1869 by a Swiss biochemist named Friederich Miescher.[16] That mysterious sticky stuff—DNA—was enough *all by itself* to transform the ordinary R into deadly R. In modern language, what made transformed R deadly was genetic material incorporated from S-strain DNA.

The bottom line? Scientists could now point to the material basis of heredity, to Mendel's factors, to genes. But rather than being made of some special kind of protein, genes were made of DNA (deoxyribonucleic acid). To find out more about genes, then, scientists would clearly need to figure out how this molecule, DNA, worked. At that point, researchers knew relatively little about DNA. From Miescher's original discovery of the substance in 1869 (just four years after Mendel published his paper on the pea), scientists knew what DNA was made of: carbon, hydrogen, oxygen, nitrogen, and phosphorus. A decade and a half later, in 1885, German biologist Albrecht Kossel discovered that DNA included four types of alkaline (opposite of acidic) molecules known as "bases," which he named cytosine, thymine, guanine, and adenine, and which we now refer to as nucleotides.[17] But the exact composition of DNA, and how those bases related to one another, seemed to differ from one species to the next. For example, the proportion of guanine was higher in the thymus of an ox than it was in the thymus of a person.[18] Unexplained were biochemist Erwin Chargaff's 1950 "laws": The amount of cytosine always seemed to match the amount of guanine, and the amount of thymine always seemed to match the amount of adenine.[19]

With Avery's discovery, and independent confirmation from Alfred Hershey and Martha Chase that followed in 1952,[20] there was soon a race to figure out DNA's exact shape and the way that its molecules fit together. The smart money was on Linus Pauling, the world's leading authority on chemical bonds. True to biology's *Daily Racing Form,*

Pauling, who later won two Nobel Prizes, was first to publish[21]—but his hypothesis—a triple helix that nowadays can only be found in science fiction—turned out to be flawed.[22] Before Pauling could spot his own error, he was overtaken by two ambitious newcomers, a twenty-five-year-old American who had only recently finished his Ph.D. dissertation, and a thirty-something British graduate student who had yet to finish his.

I am speaking, of course, about James Watson and Francis Crick. What the famous team discovered, in February 1953 (with the help of critical X rays that were taken by Rosalind Franklin[23] and Maurice Wilkins), was that the DNA molecule was a double helix: two twisted sugar-phosphate ladders connected by rungs made up of pairs of nucleotide bases.[24] The idea of a helix wasn't new. What was new was the understanding of the way in which the bases fit together: Each individual rung was made up of a pair of "unlike" bases, either an adenine (A) and a thymine (T), or a guanine (G) and a cytosine (C). The reason that the amount of adenine correlated so well with the amount of thymine was that they always came in pairs—Chargaff's laws had been explained—and the structure of DNA finally deciphered.

"It has not escaped our notice," Watson and Crick famously wrote, "that the specific pairing we have postulated immediately suggests a possible copying mechanism for the genetic material."[25] The immediate significance of their theory was in the way it connected to Mendel's questions about heredity. An organism could resemble its parent only if Mendel's factors could be transferred from parent to child, and that, in turn, required that there be some way to make copies of the factors. DNA provided for that possibility: Information was contained in the sequence of nucleotides. The two strands of the substance could separate and serve as templates for more strands—voilà, biological Xerox.

The conception of genes as templates for proteins grew out of efforts to figure out what all those A, C, G, and T nucleotides were for. Almost immediately, a physicist named George Gamow took a first stab, guessing that the amino acids that make up a protein might somehow stick into crevices between the rungs of the DNA ladder.[26] Which

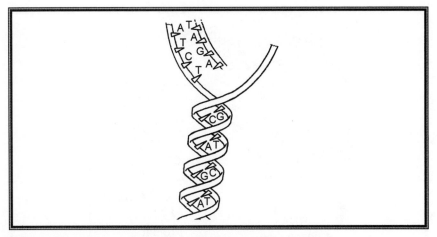

Figure 4.3   DNA, and the process of its replication
Illustration by Tim Fedak

protein emerged from a given DNA sequence would, in Gamow's theory, be a matter of which amino acids fit into the crevice between its
nucleotides. Gamow's crevice theory was wrong in its details. Proteins
are not formed through direct interaction with the DNA (and the
crevices between nucleotides are irrelevant for this process). But the
spirit of his idea was correct: One of the main ways that genes exert
their influence is by providing templates for proteins.

As became clear in the early 1960s, sequences of three nucleotides,
known as triplets, or *codons,* get translated into amino acids, with
each triplet standing for a different amino acid. For example, triplets
of T-C-G get translated into serine, triplets of G-T-T into glutamine,
and so forth, each codon serving as a template for a different amino
acid. Series of triplets were translated into the chains of amino acids,
which in turn fold up into the complex three-dimensional molecules
we know as proteins. (Aficionados will realize that I'm oversimplifying in several ways. DNA must first be copied, or "transcribed," onto
RNA (ribonucleic acid), an intermediate complement of DNA, before it gets translated into amino acids. Furthermore, there are sixty-
four codons but only twenty amino acids, so sometimes two, three, or
even six different codons serve as templates for a single amino acid;
for our purposes, these details will not matter.)

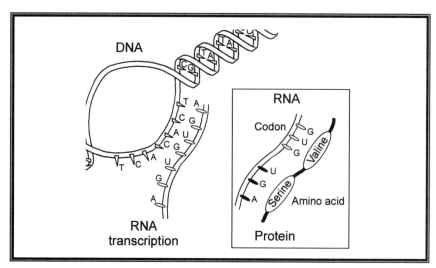

Figure 4.4   From DNA to RNA to protein
Illustration by Tim Fedak

The conception of genes as protein templates is partly correct, and it's what many people think of when they think of genes. Genes genuinely do provide templates for protein building, and many disorders—mental and physical—are the result of small "errors" in protein templates. Sickle-cell anemia, for example, is caused by a single spelling error in the 861-nucleotide-long gene for building hemoglobin, the four-part protein that allows red blood cells to carry oxygen.[27] Blood cells with ordinary hemoglobin look like dimpled discs; sickled blood cells are so named because when they are not bearing oxygen they form a crescent, or sickle, shape, the direct consequence of change in a single nucleotide from an A to a T. Sickled cells have a front and a back, and when they lack oxygen they tend to fit together like the top and bottom of a Lego brick, forming chains that can clog arteries and block the flow of blood. Most of the time, such cells commit suicide, but occasionally the clots persist and one of the body's organs may be damaged, sometimes leading to death. (Like many disorders, sickle-cell anemia is most severe in people who inherit two copies of the errant gene, one from each parent. When only a single copy is inherited, the normal copy can do some of the work and the illness is far less severe. The disorder may persist in the population because even a single copy of the "mutant" gene conveys a resistance to malaria.)

Not all genetically influenced disorders, however, can be traced to errors in protein templates. Even the Protein Template Theory was incomplete in a significant way. Proteins are marvelous molecular machines, but what makes one animal different from the next is not just its set of proteins, but the *arrangement* of those proteins, and, remarkably, the arrangements, too, are a product of the genes. The Protein Template Theory captured only half the real story. Each gene actually has two parts: the protein template, which is widely known, and a second part that provides *regulatory* information about *when that template should be used.*

This final, crucial insight—that genes provide not just templates but also instructions for regulating when a gene should be translated into protein—came in 1961, in Jacques Monod and François Jacob's investigations of the eating habits of the bacterium *Escherichia coli*.[28] These insights led scientists to refine the Protein Template Theory into the theory of genes that is now considered correct, which I will call the Autonomous Agent Theory.[29] Monod and Jacob's study began with the observation that *E. coli* could quickly switch from a diet of glucose (its preferred sugar) to one of lactose (the sugar found in milk). In ordinary glucose-rich environments, *E. coli* doesn't bother to make enzymes for metabolizing lactose. But when glucose becomes scarce, the bacteria switch their diet in just a few minutes. To manage that, they must produce thousands of copies of enzymes, such as ß-galactosidase, a molecule that facilitates the breakdown of lactose to galactose and glucose.

What Monod and Jacob discovered is that the genes for these lactose enzymes switched on or off *as needed* according to a simple logical system. The templates for lactose-metabolizing enzymes are translated into proteins if and only if exactly two things are true. First, the bacterium *must* have lactose around, and second, the bacterium must *not* have access to glucose. The logical juxtaposition of these two requirements (IF lactose AND NOT glucose) should instantly ring a bell with any reader who has computer programming experience—for the requirement "IF X AND NOT Y" is of a piece with the billions of IF-THEN rules that guide the world's software. What Jacob and Monod had discovered, in essence, was that each gene acts like a single line in a computer program.

The net result is a kind of mass empowerment: Every gene is a free agent authorized to act on its own, hence the Autonomous Agent Theory. As soon as the IF part of a gene's IF-THEN rule is satisfied, the process of translating the template part of a gene into its corresponding protein commences. There is no form to fill out in triplicate, no waiting for approval. Patrick Bateson and Richard Dawkins have described the genome as a whole (the collection of all the genes in a given organism) as a recipe, but it is also possible to think of each individual gene as a recipe for a particular protein; on the latter analogy, what IF-THEN gene regulation means is that *each recipe can act on its own.*

IF-THEN

Understanding how genomes contribute to the construction of body and brain is thus a matter of understanding how the two parts of every gene—the regulatory IF and the protein template THEN—work together to guide the fates of individual cells. Nearly every cell contains a complete copy of the genome (which is why one can grow a carrot from a clipping or clone a sheep from a single cell[30]). But most cells specialize for particular tasks, some signing up for service in the circulatory system, others in the digestive tract or the nervous system, relocating and even committing suicide when their job requires it. And it is from that process of specialization, in the individual decisions of the trillions of cells that make up a body, in how cells spend their lives, in how they grow, slip, slide, divide, and differentiate, that the structure of the body and brain emerge. (Or fail to emerge, for most birth defects stem in one way or another from errors in these basic processes.)

What makes one cell different from the next is not which genes it has copies of, but rather *which of those genes are switched on.* The recipe for hemoglobin is followed only in red blood precursors, the recipe for human growth hormone only in the pituitary gland. Some genes are expressed only in the brain, others only in the kidneys or the liver, or in a particular kind of cell, or in a particular place within a cell, and many genes are just as choosy about *when* they are expressed as they are about *where* they are expressed. "Housekeeping" genes, such as those that build proteins that help convert sugar to energy, are on almost all the time in almost every cell,[31] but most are on (or most active) only at select times, during particular situations (for example, during cell division or gastrointestinal inflammation), or at particular moments in embryological development (such as during the leg-growing, tail-shedding process of tadpole-to-frog metamorphosis).[32] In this way, by switching on only at specific times and places, genes modulate the growth of proteins in different ways in different cells. With IFS that are tied to particular times and types of cells, each cell can develop in its own unique way.

What drives the embryo forward in development—and what drives a monkey embryo to become a monkey rather than a grapefruit—is each species' unique set of IF-THENS and the different ways in which they drive cells to develop and specialize. If genes are like lines in a computer program—an IF that controls when a gene will be expressed, a THEN that says what protein it will build if it is expressed—they are a special kind of computer program, one that is followed not by a central processor but autonomously, by individual genes in individual cells.

With one more trick—*regulatory proteins* that control the expression of other genes—nature is able to tie the whole genetic system together, allowing gangs of otherwise unruly free-agent genes to come together in exquisite harmony. Rather than acting in absolute isolation, most genes act as parts of elaborate networks in which the expression of one gene is a precondition for the expression of the next. The THEN of one gene can satisfy the IF of another and thus induce it to turn on. In this way, a single gene that is at the top of a complex network can indirectly launch a cascade of hundreds or thousands of others, leading to, for example, the development of an eye or a limb.

In the words of Swiss biologist Walter Gehring, such genes can serve as "master control genes" that exert enormous power in a growing system. *Pax6,* for example, is a regulatory protein that plays a role in eye development, and Gehring has shown that artificially activating that one gene in the right spot on a fruit fly's antenna can lead to an extra eye, right there on the fly's antenna—a simple regulatory protein IF that leads, directly and indirectly, to the expression of approximately 2,500 other genes.

The IFS and THENS can even lead a single organism to develop in different ways in different circumstances. The African butterfly *Bicyclus anyana,* for instance, comes in two different forms depending on the season, a colorful wet-season form, and a duller brown dry-season form. Which one develops is determined only late in the larval stage of the butterfly's development, probably on the basis of a temperature-sensitive gene that triggers different cascades depending on the climate. Genetically identical butterflies raised in a warm laboratory tend to take on the wet-season form, whereas those raised in cooler temperatures tend to take on the dry-season form. It would be impossible for the genome to "know" in advance whether a particular larva will develop in the wet season or the dry season, so instead nature has endowed the *B. anyana* genome with IF-THEN instructions for handling both, and machinery for letting the environment determine the most appropriate phenotype.[33]

HISTORY AND GEOGRAPHY

In simple organisms, many of the IF-THEN cascades of development are driven primarily by a cell's history. The growth of the *Caenorhabditis elegans* roundworm is so regular that biologists have taken to drawing "fate maps," or "lineages," diagrams that would make a genealogist feel at home. Each newly fertilized egg divides four times, each time budding off from a different founder cell, each of which, under normal circumstances, has a specific destiny. For example, the founder cell known as "D" generally gives rise to muscle cells, and founder cell "AB" generally gives rise to neural cells, muscle cells, and a set of "hypodermal cells" that lie in a layer beneath the skin's surface. The first few generations are shown in the first figure here. By the time all the great-great-great-grandchildren are born, the chart is a lot more complicated, but it still looks like a family tree, as in the second figure.

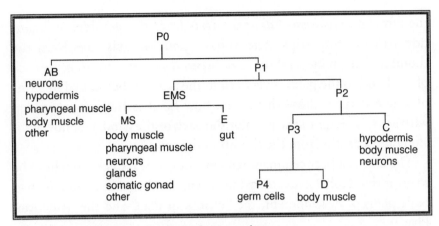

Figure 4.5   Cell fate in an early *C. elegans* embryo

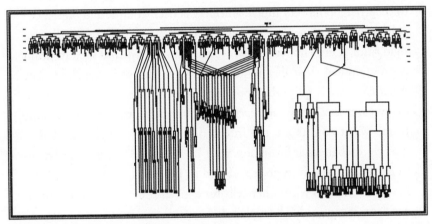

Figure 4.6   Further development in *C. elegans*

In the nematode, most cells appear almost as if they were on auto-pilot, going about their business independently, according to a strict schedule, even if, say, the head is missing. In mammals, cells do use a bit of history, but they also rely heavily on molecular signposts that tell growing cells where they are. In complex, three-dimensional biological structures like the human heart or arm, the body makes use of at least three different systems, or *axes* (plural of "axis"), for indicating position, each depending on a different set of genes and proteins. In the arm, the *proximal-distal* axis starts from the shoulder and runs down to the fingertips. The *anterior-posterior* axis runs from the thumb to the pinkie, and the *dorsal-ventral* system runs from the back of the hand to the palm. Every cell in the arm can be defined in terms of where it is on those three axes: how far it is from the shoulder, whether it is closer to the palm or the back of the hand, and how far it along the axis it is from the thumb to the little finger.

The contrast between mammals and worms was once thought to be so great that Nobel laureate Sydney Brenner joked that there were two basic plans of development, the "European Plan" and the "American Plan":

> The European way is for the cells to do their own thing and not to talk to their neighbors very much. Ancestry is what counts, and once a cell is born in a certain place it will stay there and develop according to rigid rules; it does not care about the neighborhood, and even its death is programmed. If it dies in an accident, it cannot be replaced. The American way is quite the opposite. Ancestry does not count, and in many cases a cell may not even know its ancestors or where it came from. What counts are the interactions with its neighbors. It frequently exchanges information with its fellow cells and often has to move to accomplish its goals and find its proper place. It is quite flexible and competes with other cells for a given function. If it dies in an accident, it can readily be replaced.[34]

In truth, all animals make use of both kinds of information, albeit in somewhat different proportions. Despite its apparent reliance on ancestry, the hermaphroditic worm still uses positional information to, for example, configure its egg-laying opening (otherwise known as the worm's vulva). The worm's vulva generally consists of exactly twenty-two cells, which originate, under normal circumstances, at a cell known as $P6_p$, regular as clockwork. But the worm will still grow a vulva if inquisitive experimenters use a laser to destroy $P6_p$. As developmental biologist Judith Kimble discovered, there are actually six skin cells that have the potential to give rise to the vulva. Which one actually does so is determined not by a blueprint but by a protein signal that is secreted from a cell known as the "anchor cell."[35] The skin cell closest to the anchor cell then gives rise to the primary vulva cells, while the two adjacent skin cells become secondary vulva cells. If the anchor cell is destroyed (by the zap of the laser beam), no vulva grows. If the anchor cell is shifted toward the head, the vulva shifts in the same direction, centering around the anchor cell's new position rather than around its ordinary position. What triggers the vulva program is thus not an absolute cue to location but a functional one, a triggering of a receptor for a cue given off by the anchor cell.

In mammals, many of a cell's IF-THEN decisions depend on a mix of ancestry and signals. One of the first studies to pit the two against each other came in the 1950s. Embryologist John Saunders Jr. took presumptive thigh tissue (tissue that would ordinarily turn into a thigh) from a chicken embryo and implanted it onto the edge of the wing bud of another chick embryo. The transplanted tissue didn't simply fill in the missing part of the wing tip, but it didn't turn into a thigh, either. Neither ancestry nor neighborhood won out. Instead, claws sprouted from the ends of the chicken's wings.[36] The transplanted tissue retained a memory (in the form of molecular markers) of its lineage (from the leg) and combined that with the positional cues from its new environment (in the edge of the wing bud), rendering dramatic the complex calculus of combining position and ancestry. And it is that same calculus that allows presumptive eye cells to become stomach cells and presumptive somatosensory cells to become visual cells: By including position in the equations that determine cell fate, the growing mammal automatically achieves a large degree of flexibility.

What propels an embryo from one stage to the next—and makes one species different from another—is not a blueprint but rather an enormous autonomous library of the instructions contained within its genome. Each gene does double duty, specifying both a recipe for a protein and a set of regulatory conditions for when and where it should be built. Taken together, suites of these IF-THEN genes give cells the power to act as parts of complicated improvisational orchestras. Like real musicians, what they play depends on both their own artistic impulses and what the other members of the orchestra are playing. As we will see in the next chapter, every bit of this process—from the Cellular Big 4 to the combination of regulatory cues—holds as much for development of the brain as it does for the body.

How much can you do with a system like that? Consider the power of groups of simpleminded ants that work together to build a colony. Outside of DreamWorks Studios, individual ants can do little more than follow one chemical trail or another, pretty much insensitive to the rest of the world around them, yet their collective action yields great complexity.

In a similar way, individual genes are not particularly clever—this one only cares about that molecule, that one only about some other molecule. The regulatory region that controls insulin production, for example, looks for signs that it is in the pancreas, but it can easily be fooled. It's not smart enough to look around and realize that it might the victim of a party prank played in a Petri dish. But that simplicity is no barrier to building enormous complexity. If you can build an ant colony with just a few different kinds of simpleminded ants (workers, drones, and the like), think what you can do with 30,000 cascading genes, deployed at will.

5

COPERNICUS'S
REVENGE

Finally we shall place the Sun himself at the center of the Universe.
All this is suggested by the systematic procession of events and the harmony of the whole Universe, if only we face the facts, as they say, "with
both eyes open."

—Copernicus

The human brain has variously been described as "the last frontier,"[1]
"biology's greatest challenge,"[2] "the most elaborate structure in the
known universe,"[3] and Woody Allen's "second favorite organ."

In some ways, the brain seems unlike the rest of the body. Sure, it
depends on the flow of blood and oxygen just like the rest of the body,
but only the brain *thinks*. Because the brain is the physical realization
of mental life, the root of language, mathematics, and emotion, it is
tempting to think that its origins are somehow different. If we are not,
as Copernicus showed, at the center of the universe, at least, surely,
there must be something awfully special about our brains. Physician
Richard Restak wrote, "Since the brain is unlike any other structure in
the known universe, it seems reasonable to expect that our understanding of its functioning—if it can ever be achieved—will require
approaches that are drastically different from the way we understand
other physical systems."[4]

The notion that the brain is drastically different from other physical systems has a long tradition; it can be seen as a modernized version of the ancient belief that the mind and body were wholly separate. But the past 150 years have made it amply clear that the brain is a physical system and that changes to the brain lead to changes in the mind. Although the function of the brain is different from that of other organs, the brain's capabilities, like those of other organs, emerge from its physical properties. We now know that strokes and gunshot wounds can interfere with language by destroying parts of the brain, and we know that drugs such as Prozac and Ritalin can influence mood by altering the flow of neurotransmitters.[5] The fundamental components of the brain—the neurons and the synapses that connect them—can be understood as physical systems with chemical and electrical properties that follow from their composition.[6] As we will see, the IFS and THENS of genes guiding the growth of neurons in much the same way they guide the growth of any other type of cell. In many ways, the development of the brain is simply a special case of the development of the body.

∞

William Shakespeare wrote that "we are such stuff as dreams are made on," and Herman Melville's great-great-grand-nephew (the pop star known as "Moby") sang that "we are all made of stars." But, in reality, we are all made of atoms, and that holds as true for the brain as for any other organ. We are made up mainly of carbon, hydrogen, oxygen, phosphorus, potassium, nitrogen, sulfur, calcium, and iron, memorialized in the biochemist's mnemonic *C. HOPKiNS CaFe*. Those atoms, in turn, combine to make complex molecules; after water, protein and fats are the most common in living things. (Compared to your biceps, the brain has a bit more fat—much of it in the form of myelin, which insulates the "wiring" that runs between neurons—and a little less protein, but the differences are slight.[7] Puree either in a Cuisinart, and you would wind up with more or less the same soup.)

As we move up from atoms to cells, the same general point holds. All organs, including the brain, are made up primarily of cells. The special nerve cells of the brain—neurons—look, at first glance, to be rather different from most other kinds of cells. They are often (though not always) larger than most other cells, and they are flanged on one side by long *axons* that carry signals away from the cell, sometimes extending the length of the body, and on the other by treelike *dendrites* that allow neurons to receive signals from thousands of other nerve cells. Neurons are electrically alive, capable of sending brief jolts of charged atoms down the lengths of their axons, and, even more remarkably, they are smart. Not as smart as a person, but smart enough to synthesize vast arrays of information,[8] and fast enough that a group of them working together can recognize a word or a familiar object in a fifth of a second.[9]

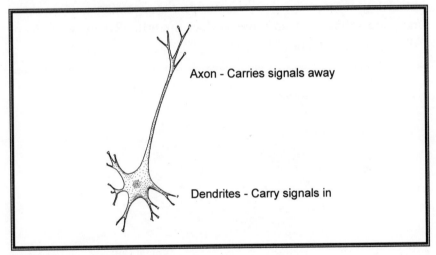

Figure 5.1   A neuron
Illustration by Tim Fedak

Despite these qualities, neurons are still just cells, better thought of not as unique creatures but as specializations of a general cellular plan that is widely shared across the body. Although their outward appearance and special talents for computation and long-distance communication make them seem quite different from most other cells, under the hood most of what neurons do is much the same as what other cells do. Their cell bodies (known as *somas*) contain essentially the same variety of micro-organs (known as *organelles*) as a skin cell or a liver cell: mitochondria to generate energy, protein synthesis plants known as endoplasmic reticula, membranes to keep invaders out, and nuclei to keep the DNA in. Any given neuron, in fact, begins its life as an epithelial cell, and but for the grace of a few chemical cues, it could just as easily wind up on the outside, as a skin cell.

Many of a neuron's most spectacular specializations are just variations on ordinary cellular themes.[10] It has more mitochondria than usual, for example, so as to support its high demands for energy. Even the long, spindly axons aren't something wholly new. The fibrous cytoskeleton proteins that axons rely on for structure, and the track-like microtubules[11] that they rely on for transporting materials, are both found in virtually all cells. Neurons, the characteristic cells of the brain, are special, but no more so than the other 210 or so kinds of cells in the human body.[12]

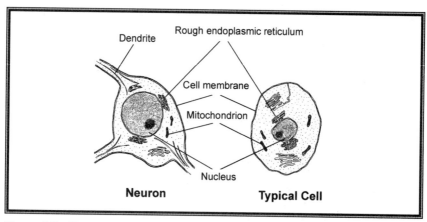

Figure 5.2   How neurons resemble other cells
Illustration by Tim Fedak

THE MAKING OF A BRAIN

The process nature uses in building a brain very much resembles the process it uses in building many other organs. At the broadest level, brain building involves the same kind of successive approximation that Aristotle identified. The brain comes into being not all at once, but in a series of stages, first as a simple group of undifferentiated cells that soon thicken and curl to form first a sheet and then a tube. That tube sprouts more bulges, and those bulges sprout bulges, each of which is further refined. One divides into a series of segments that collectively compose your hindbrain, an evolutionarily ancient command center of nerves that contributes to processes such as respiration, balance, and alertness. A second gives rise to the midbrain, which coordinates visual and auditory reflexes and controls functions such as eye movements. The surface of another bulge gives rise to the precursor of your forebrain, vital to decision making and reason. Each section becomes more and more refined over time, bending and curling, flexing and folding, pursing and puckering, elongating and extending much like any other organ.

At a finer level, brain building depends on the same cellular-level processes—division, differentiation, migration, and planned cell death—as other organs. For example, the life of a neuron, whether found in the brain or in the spinal cord, begins with an act of cell division, and the amount of such cell division very directly drives brain size. The main reason our brains are three times bigger than chimp brains is not that we have more experience, but that we have more neural cell division.

Once brain cells are born, they must specialize and migrate to their final locations. If brain cells fail to migrate, or migrate to the wrong places, the result can be birth defects ranging from cerebral palsy to Kallman syndrome (which leads to sterility and lack of a sense of smell), or to lissencephaly (from the Greek words *lissos,* for "smooth," and *encephalos,* for "brain," a disorder in which the brain develops without its usual convolutions).[13]

Cell differentiation can turn neurons into everything from clocks that control circadian rhythms to photoreceptors that convert light into electrical-chemical impulses or decision makers that tally votes and decide courses of action. In the retina (often used as a case study because it can be directly and naturally stimulated), there are at least fifty different kinds of neurons specialized to different tasks, such as looking for motion, recognizing colors, detecting objects in low light, and measuring brightness and contrast.[14] In the brain as a whole, there may be as many as 100,000 different kinds of neurons, each contributing to a different aspect of mental life.[15] (Here, too, there is always the chance for error; some congenital muscular dystrophies, for example, may stem from errors in the process by which motor neurons—the neurons that drive muscle cells—take on their fates.)

Programmed cell death—deliberate cellular suicide—helps to fine-tune particular populations of neurons. Many aspects of brain development, such as matching sensory areas of the brain to inputs from elsewhere in the body, seem to rely on a two-part strategy. The body initially builds more cells in particular brain regions than it ultimately requires, and then, in a kind of Darwinian fight-to-the-death strategy, culls those that fail to integrate into some larger system.[16] Like anything else in development, the process of cell death must be tuned just so: too much, and there are too few cells left to do the job; too little, and some unnecessary hangers-on get in the way. The process is extremely sensitive to drugs, ranging from anesthetics to alcohol, which is one of the main reasons that pregnant women are strongly advised to watch what they ingest.[17] The choices of a cell in the brain are not so different from those facing the cells in the rest of the body.

Each of these cellular processes—division, migration, differentiation, and planned cell death—what I like to think of as the Cellular Big 4—is quite intricate. Nerve cell migration, for example, proceeds in roughly four steps. First, some cellular system has to give the neuron in question a "green light"; second, the neuron has to figure out where to go; third, it has to engage its "motor"; and finally, it has to know when to stop. Each neuron (or group of neurons) has its own set of instructions. Many neurons in the shell-like cerebral cortex, for instance, originate on the inside surface of the shell and then gradually climb (along a set of pole-like "radial glia cells") toward the outer surface of the shell, moving further outward the later they are born.[18] Other cells, "born" in different places, shift (tangentially) along the surface of the shell.[19] Even in a simple organism like a worm, the mechanics of migration are so complicated they could have been borrowed from one of John Madden's football playbooks. Cell number 1 goes right, number 2 goes left, and cell 3 goes long for a pass.[20]

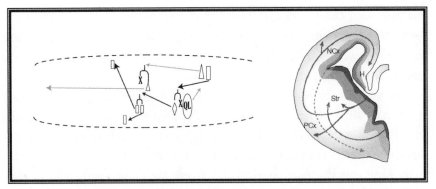

Figure 5.3 Migration pathways in the development of the nervous system of a
worm (left) and a mouse (right).
SOURCE: C. Kenyon, by permission of the author and the company of biologists J. Rubenstein and O. Marin, by permission of the author and *Nature Neuroscience*

To a remarkably large extent, all this complexity is guided by gene
expression. Indeed, without gene expression, there would be no migration, no differentiation, no division, no planned cell death—mul-
ticellular life as we know it would not exist.[21] Starting in the
mid-1990s, developmental neuroscientists such as John Rubenstein
and Christopher Walsh finally began to work out the genetic "codes"
that regulate these processes, specific genes that affect the ways in
which the Cellular Big 4 help to sculpt the brain. By flipping the right
genetic switches, these researchers have been able to grow mice with
abnormally large brains by inducing extra cell division,[22] trick differ-
entiating neurons that would ordinarily produce excitatory neurotransmitters into producing inhibitory neurotransmitters (the neural
equivalent of getting Democrats to vote Republican),[23] and coax neurons that would otherwise be bound for the cortex to instead head un-
derground, to a subcortical area known as the *striatum*.[24]

The point of such experiments, of course, is not simply to show off
the power of advances in biotechnology, but rather to work out the
precise role of genes in guiding brain development. The upshot of
the hundreds of experiments—most conducted since the beginning
of the new millennium—is this: Genes guide neural development in
precise and powerful ways, modulating virtually every process that is
important in the life of a cell, by controlling the production of the

enzymes and cellular components that give neurons their shape and form, by controlling the placement and guidance of the motors that move those cells, and by issuing the commands that, when necessary, lead to their death.

The regulatory regions that direct those genes are guided in no small part by an intricate system of signposts and landmarks, made mostly of highly specialized signaling proteins. Such proteins (as always, the products of genes) often act a bit like radio waves that gradually fade out the further they get from their source. In the body, because they decrease gradually as they move away from the source, such signals are known as *gradients*.

For example, the gradient of a protein known as *"FGF8"* serves as a cue for the development of "barrel fields"—sets of cortical neurons in rats and mice that respond to whisker stimulation. Under ordinary circumstances, *FGF8* is concentrated most heavily toward the front of the brain, least heavily toward the back. Artificially altering that gradient profoundly alters the placement of the mouse's barrel fields. If the concentration of the *FGF8* signal is increased early in development, the barrel fields grow unusually far forward in the brain. If the concentration of the *FGF8* signal is decreased, the barrel fields grow unusually far back. The most amazing result came in 2001 when developmental neuroscientists Elizabeth Grove and Tomomi Fukuchi-Shimogori placed an extra bead of *FGF8* opposite its usual location, so that the two concentrations of the protein would be like two radio stations broadcasting the same signal at opposite ends of a valley. The result? The double set yielded two gradients of *FGF8* and a double set of barrel fields, one the mirror image of the other.[25]

Dozens of other signal-beacon genes broadcast, to extend the radio analogy, on different frequencies, each influencing a different aspect of neural development. For example, *FGF8* works in part by controlling a gene called *"Emx2."*[26] Disabling the *Emx2* genes in a growing mouse changes the relative balance between the hippocampus and the frontal cortex, squashing the hippocampus to make room for more frontal matter—one more way in which the subtle specializations of the brain can be profoundly altered by changes in even a single gene.[27]

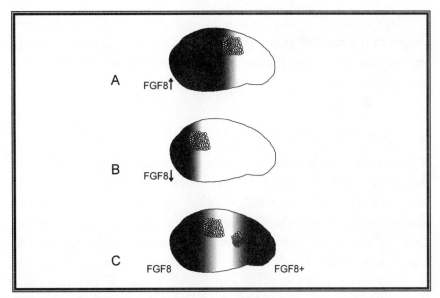

A FGF8↑

B FGF8↓

C FGF8   FGF8+

Figure 5.4   Gradients guide neural development
Illustration by Tim Fedak

What's amazing is how little of the overall scheme for embryonic development is special to the brain. Although thousands of genes are involved in brain development, a large number of them are shared with (or have close counterparts in) genes that guide the development of the rest of the body. The "motors," for example, that allow neurons to move depend on a special protein called actin that can contract so as to pull the back edge of the cell forward toward the leading edge— exactly what actin does in limb development as it pulls finger cells toward the hand and toe cells toward the foot.[28]

More generally, around 500 "housekeeping genes"—genes that guide processes such as metabolism, cell death, and the synthesis of proteins—do essentially the same things in the brain as they do elsewhere.[29] *"Ced3"* and *"ced4,"* for example, lead to cell death in the brain just as they do in the liver and in the webbing between an embryo's fingers, and insulin plays much the same role in the metabolism of glucose in the brain as it does in the rest of the body.[30] Scores of others belong to protein "families" that have ancestors used elsewhere in the body.[31]

Scientists actually have to look fairly hard to find genes that are *unique* to the brain,[32] and as we will see in Chapter 7, many of those are just variants on old themes, new arrangements of old proteins in more precise ways.

MENDEL'S MIND

If genes influence the development of the brain, do they also influence the development of mind and behavior? In the animal world, the answer is clearly yes. Since the late 1990s, developmental neuroscientists—using new techniques that allow them (among other things) to disrupt ("knock out") or alter particular genes[33]—have made great advances, moving from early studies that tried to show that genes matter at all (for example, by showing that breeding can affect certain traits or that, à la human twin studies, closely related animals are more similar than distantly related ones) to highly focused studies that have tied particular traits to particular genes.

For example, one set of studies has looked at foraging habits in worms *(Caenorhabditis elegans)* and fruit flies *(Drosophila melanogaster)*. *C. elegans* worms, as it happens, feed on *E. coli,* the very bacterium that led Monod and Jacob's discovery of gene regulation. Some *C. elegans* prefer to forage in groups; others are loners. This difference in worm behavior has been traced to a difference in a single amino acid in the protein template region of a gene known as *npr1:* Worms with the amino acid valine in the critical spot are "social," and worms with phenylalanine are loners. In flies, a gene called *for* (for "foraging") controls a distinction between a group of flies called "rovers," which wander from food patch to food patch, and a group of "sitters," which tend to stick within a particular food patch.[34] (Sitters aren't simply sluggish—when there is no food around, they scurry around at the same speed as rovers.)

Showing a correlation, however, is one thing, and showing that a particular gene actually *causes* a change in behavior is something else, even more impressive. In the past decade, development neuroscientists have devised literally dozens of studies that do exactly that. For example, Marla Sokolowski of the University of Toronto has shown how one can actually switch a strain of the loner *C. elegans* worms into social worms simply by altering their genomes to include the valine version of the relevant gene. Larry Young and Tom Insel at Emory University showed how changing the regulatory IF region of a single gene could have a large effect on the social behavior of mammals. After observing that differences in sociability in different species of voles was correlated with how many vasopressin receptors they had, Young and Insel transferred the regulatory IF region of sociable prairie voles' vasopressin gene into the genome of a mouse—and in so doing created mutant mice that were more social than normal.[35]

A team of National Institutes of Health (NIH) researchers led by Dennis Murphy created a strain of anxious, fearful mice by disabling a gene that produces a protein that transports serotonin.[36] Five other labs have shown that disrupting a gene that produces serotonin *receptors* also leads to mice with increased fear and anxiety.[37] By disrupting a gene known as *"Hoxb8,"* Joy Greer and Mario Capecchi at the University of Utah created a mouse that groomed itself constantly, pulling and tugging on its own hair until it was bald.[38] And The Mice that Groom Themselves Too Much are just the tip of the iceberg. To take but a few examples, so-called knockout techniques have also produced mice that lack the nurturing instinct,[39] "hyperactive" mice,[40] hypersensitive mice that are especially reactive to stress,[41] and mice that progressively increase alcohol consumption under stress.[42]

Breeding studies show that genes matter, and correlational studies tie particular bits of the genetic code to particular changes in behavior. The most recent studies have gone a step further, showing how deliberately altering genomes can directly alter behavior.

∞

We cannot, of course, ethically alter genomes to study the effect of genes on the human mind, but at least three lines of evidence strongly suggest that genes play much the same role in the development of the human mind as they do in the animal mind.

One argument comes from disorders of mind and behavior. Although many, perhaps most, disorders of the mind cannot yet be directly tied to disruptions in particular genes (for reasons I discuss in the Appendix), many can, and there is no longer room for serious doubt about the extent to which naturally occurring alterations to the genome can alter the human mind. An NIH website catalogs literally thousands of disorders that can be linked to alterations in single genes. Many of these directly impair development of the brain, including not only PKU and several forms of lissencephaly but also Huntington's disease (the neurogenerative disease that folksinger Woody Guthrie died from), Angelman syndrome (once referred to as "happy puppet" syndrome, a disorder that causes a happy disposition accompanied by severe retardation and unusual facial expressions), certain forms of Alzheimer's disease and Parkinson's disease, and a rare speech and language disorder that I will discuss in Chapter 7.[43] (In so-called "single-gene disorders," the responsible gene is better thought of as a broken link in a complex chain, not something that by itself is wholly sufficient for the entire series of events in the neurons that leads to a particular behavior. In even more complex disorders, such as dyslexia, autism, and schizophrenia, symptoms may depend on subtle, difficult-to-detect interactions between multiple genes—as well as the environment—but there is ample evidence that genes play an important role in the origins of even these disorders.[44])

Alterations to the genome can also lead to differences between normal individuals: A small but growing literature has been able to tie human individual differences to particular genetic loci. A 2003 study tied people's talents for remembering events to the specific version of a particular nerve growth protein they possessed; those with the amino acid valine in the relevant position significantly outscored those with methionine.[45] Another study (that I will discuss in more depth later) found that people with a certain version of a gene that codes for an enzyme involved in the process of metabolizing neurotransmitters such as serotonin and dopamine are—under specific environmental conditions—at a greater risk for committing violence,[46] and a third (echoing a mouse study I mentioned earlier) showed a link between anxiety and a gene for a protein that transports serotonin.[47]

But these data are just starting to come in. To my mind, the strongest argument for a link between genes and the human mind comes from the studies of animals. Earlier in the chapter, I observed that most of the genes in the brain are related to genes expressed elsewhere in the body; few genes are entirely new. The corollary is that virtually every gene expressed in the human brain is also expressed (or closely related to a gene that is expressed) in the brain of a mouse.[48]

If the human genome were sui generis, entirely different from any other animal's genome, we probably wouldn't know much at all about the likely influence of genes upon the mind. But to an astonishing degree, what's sauce for the goose is sauce for the gander. Virtually every gene in a mouse genome[49]—and many in the genome of a fruit fly[50]— has some sort of counterpart in the human genome. As we will see in Chapter 7, evolution didn't start from scratch when it built the human brain. Many of the genes and proteins that participate in the construction of the brain have histories that date back to a time long before primates branched off from other mammals; some can be traced all the way back to bacteria. Which means that when scientists work out the function of a gene in an animal—say, a mouse or a fly—they are on solid ground to suspect that the gene may do something quite similar in a human. *Pax6*, for example, doesn't just guide eye formation in flies; it is also critical for eye formation in both humans and mice.[51] As we saw, serotonin modulates anxiety in mice, and it does the same thing in humans. Animals with deletions of the *Fmr1* gene have disturbances in a variety of different brain areas; humans with the comparable deletion suffer from a severe form of retardation known as Fragile-X syndrome.[52] We can't do the same experiments in people as we can in animals, but it is a very good bet that the general picture would be the same.[53] The mind, like the body, is significantly influenced by genes.

WHAT'S IN A "MENTAL GENE"?

Although the mind is significantly influenced by genes, it is not fixed by the genes—recall the difference between rigid hardwiring and flexible prewiring—and the connection between genes and the mind is far less straightforward than scientists might have hoped. The genes for building the mind are no more blueprint-like than the genes for building the body, and the relationship between genes and mental traits is at least as complex as the relationship between genes and physical traits. In the Matt Davies cartoon that appears in Chapter 4, individual genes (actually individual DNA nucleotides) stood for specific traits—"delusions of stock market savvy," for instance, or the "propensity to talk about the weather." Real genes really could *influence* such traits—I'll explain how in a moment—but they couldn't possibly be solely responsible for them. It is highly unlikely that any single gene would ever be solely responsible for an entire complex behavior.

In fact, I use the term "mental gene" as a bit of a joke. Although many genes affect our mental life—how we perceive and think about the world—"mental genes" are pretty much the same as other genes: self-regulated instructions for building parts of a very complex biological structure. The genes that build the mind (or at least its more tangible proxy, the brain) do much the same thing as other genes, and indeed, as we have already seen, many of them (such as the housekeeping genes found in every cell) *are* the same. From the perspective of the toolkit of biology, there is little difference between a gene expressed in the brain and a gene expressed elsewhere. A gene is a gene is a gene.

And no gene works on its own. Complex biological structures—whether we speak of the brain or of hearts or kidneys—are the product of the concerted actions and interactions of many genes, not just one. One reason that it makes no sense to talk about a gene "for" a particular behavior is that the neural circuitry involved in producing any given behavior is far more complex than any one gene. There can no more be a single gene for language, or for the propensity for talking about the weather, than there can be for the left ventricle of a human heart. Even a single brain cell—or a single heart cell—is the product of many proteins and hence many genes.

And except, perhaps, in the case of reflexes, most behaviors are the product of many neural circuits. In a mammal or a bird, virtually every action depends on a coming together of a multiplicity of systems for perception, attention, motivation, and so forth. Whether or not a pigeon pecks a lever to get a pellet depends on whether it is hungry, whether it is tired, whether there is anything else more interesting around, and so forth.

Furthermore, even within a single system, genes rarely participate directly "on-line," in part because they are just too slow. Genes do seem to play an active, major role in "off-line" processing, such as consolidation of long-term memory (which can even happen during sleep,[54]) but when it comes to rapid on-line decision making, genes, which work on a time scale of seconds or minutes, turn over the reins to neurons, which act on a scale of hundredths of a second. The chief contribution of genes in the moment-by-moment actions of an animal comes in advance, in laying down and adjusting neural circuitry, not in the moment-by-moment running of the nervous system. Genes build neural structures—not behavior.

The closest I've seen to a single gene being responsible for a whole behavior is a gene known as *ELH* found in the sea slug *Aplysia*. When *ELH*'s protein product is injected into an *Aplysia,* the hermaphroditic sea slug will begin to go through all the complex steps of its natural egg-laying behavior, which commence with the five- to ten-pound creature spewing out a long string of as many as a million eggs. The egg-laying behavior continues as the slug waves its head to extract the eggs further. Eventually, the slug winds the eggs up into a solid mass, and in the finale, it jerks its head and attaches the whole glob to something solid, such as a rock. As you might have guessed, *ELH* stands for "egg-laying hormone," but what coordinates all the component actions is not just a single hormone but a special kind of protein, called a "polyprotein," that under the right circumstances gets carved up into many subcomponents. In essence, the 271-amino-acid-long protein product of *ELH* acts not as one protein but as many, including a 36-amino-acid-long hormone that acts on abdominal neurons and stimulates the egg-laying itself as well as a variety of other neurotransmitters that excite and inhibit other neurons involved in the extracting, winding, and jerking of the eggs—a single gene conducts an entire symphony of egg-laying behaviors.[55] Yet even here, in a case where a single gene has pretty broad effects, *ELH* doesn't act on its own but calls other, preexisting neural circuits into action.

The fruit fly gene *fru* provides another example of how a single gene can play an important role in mediating a behavior—in this case, not "on-line," moment-by-moment action, but developmentally, by guiding disparate aspects of brain wiring. Recall, from Chapter 2, the complexities of a male fruit fly's courtship ritual: the vibrating, the rubbing, the licking, the curling, the consummation. Through a subtle system of "alternative splicing," in which a single gene gets translated into different proteins depending on the context, *fru* seems to participate in just about every step of the ritual, contributing to every aspect of courtship from wing vibration to copulation itself.[56] Mutations to different parts of the gene can lead to everything from "gay" flies that court males to eunuchs that don't have sex at all. And yet even *fru* cannot construct an entire circuit for behavior; like the "master control gene" *Pax6, fru*'s protein product works by guiding cascades of other events, not by acting on its own.[57]

Although there is unlikely to be any single gene for complex traits, there are likely to be many genes that profoundly influence those traits by tweaking (for better or worse) machinery that is already in place. Vasopressin, for example, the protein product of the gene that was altered in Insel and Young's hypersocial mice, is not by itself solely responsible for social behavior—it just influences the likelihood that the neural circuitry underlying social behavior (quite likely the product of other genes, not to mention environmental influences) will be invoked. Yet its effects are undeniable. If it is highly doubtful that we are born with a circuit that leads us directly to talk about the weather (or to dream of stock-market grandeur), it is not so hard to imagine that there might be genes that increase or decrease more abstract traits, such as the desire for social comfort. In this way, by influencing such more abstract traits (e.g., by modulating the synthesis or trafficking of specific neurotransmitters), single genes can have an important, albeit indirect, influence on very specific traits. If it turns out that identical twins are more likely to share propensities for weather-chatting or stock-market gambling than nonidentical twins (a good bet, given the massively consistent results of heritability studies), it will not be because there is *a* gene for talking about the weather or playing the market, but because those specific tendencies depend on a great many interacting genes that influence our different needs, desires, interests, and talents—and identical twins happen to share them all.[58]

An analogy that's not half bad is with the construction of a car. If cars were built by genes, there might be genes for synthesizing different kinds of raw materials (rubber, fabric, and steel), genes for supervising the construction of subassemblies, and genes for indicating locations within the self-assembling car. Mutations to individual genes might cause problems—an error in the recipe for creating rubber might cause many rubber parts to dissolve more quickly than anticipated, or an error in genes for connecting the spark plugs might lead to an "embryonic lethal mutation" in which the car could never leave the factory. There would be no one "steering gene," nor "propulsion gene," yet thousands of genes might affect these systems in more or less direct ways. Steering, for example, could be influenced by something as direct as genes involved in the construction of the steering column or the rack-and-pinion that transmits commands from the steering wheel to the axles, or by something as indirect as genes involved in synthesizing the rubber of the tires. We should expect things to be no less simple when it comes to the brain—no one gene for building the neural circuitry for language, for decision making, or for perception; no one way in which systems can go wrong. (In fact, since the brain is the product of a random, foresightless process of natural selection rather than a neat process of engineering, the brain is likely to be even harder to understand.)

All this is, I'm afraid, bad news for anyone who might be hoping for a quick fix when it comes to mental disorders. Anyone expecting that we will soon come to understand all there is to know about congenital disorders should reflect for a moment on how hard it can be to diagnose, say, an electrical problem in a car, even when the complete schematic is available from the factory. When newspapers report the discovery of a gene "for" obesity, "for" alcoholism, or "for" language, always remember that the discovered gene is just one among many.

There could, for example, be hundreds of different genes that contribute to obesity, influence metabolism and hunger, or regulate more abstract things such as the neural circuitry that underlies mood. Disorders such as autism and specific types of language impairment may stem from master control genes gone awry, but they could also stem from any broken link in the chain. We should not expect, in fact, that every congenital language impairment (for example) is caused by the same genetic mutation, or that if we discover the molecular basis of one language disorder that we will immediately understand them all. We should expect to find many different language impairments, each stemming from a different broken link. Disorders simply are not going to be easy to understand (more about why in the Appendix), because the connection between any given set of symptoms, what a biologist might call a "phenotype," and the underlying genes (the "genotype") is almost unimaginably complex. (All the more so, since we don't have the benefit of working from a schematic.)

But we shouldn't mistake this bad news—the complexity of the relationship between genes and finished product and the difficulty in readily understanding the causes of disorders—for a notion that nature has nothing to say. That the relationship between genes and brains (or minds) is complex does not mean that it is irrelevant. Critics of the idea that there might be "innate" mental structure have suggested that because there has thus far been no smoking gun—no single gene that has been linked to language and only language—we should abandon the idea that there is a built-in language "instinct." Others have gone further, criticizing the whole idea that the mind and brain might consist of a set of specialized modules—that the mind might be like a Swiss Army Knife. The view that I advocated in Chapter 2—that we are born with a whole slew of specialized mechanisms (including some for different kinds of learning)—has been criticized as being biologically implausible because, in the 1998 words of British psychologist Annette Karmiloff-Smith, "so far no gene . . . has been identified that is expressed solely in a specific region of cortex."[59]

But the unique marker argument is a red herring. A half-decade after Karmiloff-Smith made her suggestion, it does still seem to be the case that few genes are uniquely expressed in a single cortical area. (One study published in 2000 compared the expression of 10,000 gene fragments in the front and back of the neocortex of sixteen-day-old rat embryos and didn't find a single one that was restricted just to the front or just to the back.[60]) But the whole argument radically underestimates Mother Nature. More recent evidence has shown that genes do in fact help specify the fates of particular cortical areas—in subtle ways that do not rely on *unique* protein markers.

An alternative way of specifying the fates of particular cortical areas relies on gradients—those signaling molecules that diminish in concentration as they move away from their source. For example, a gradient of *Emx2*, one of the cortical patterning genes that is shared by all vertebrate animals, might play a strong role in guiding boundaries between sensory and motor areas. Rather than being discretely restricted to specific cortical areas, the protein product of *Emx2* diminishes gradually from a source at the back of the cortex, so it is not uniquely expressed in any particular region. But it is *differently* expressed, and that's enough of a guide to dramatically affect development. Knocking it out dramatically shifts the boundaries between cortical areas—and perhaps enough to allow different areas to specialize in different ways.

Another way in which different cortical areas could emerge without relying on a unique molecular marker for each area is that particular areas could be specified by means of *combinations* of overlapping molecular markers. On a chessboard, for example, each square can be specified by a combination of two bits of information: its rank (first row, second row, etc.) and its file (first column, second column, etc.). The white king starts at *e1*, the black king at *e8*, and so forth. It's long been clear that the body must do something similar,[61] and the most recent studies suggest that the brain does, too.[62] For example, whether a given brain-cell-in-training heads to V1 (a brain area that is important for visual processing) depends not on any single, special "V1 marker," but on a *set* of markers that work together: a perfectly fine way for a genome to help specify the complex neural structure that is presumably necessary for a complex mind.

These two "tricks"—combinatorial cues and segmentation by gradually sharpening gradients—are hardly new. Although scientists are only just discovering their importance in the brain, from the perspective of evolution, both cues are ancient. The gradients, for example, play a critical role in the segmentation of a growing fruit fly's larvae, and combinatorial cues play a critical role in the development of a fly's eye.[63] Both provide ways of genetically inducing different parts of the brain (or body) to take on different functions. Nature has been in the business of building biological structures for an awfully long time, enough to dope out many of the best tricks, and, as we will see, stingy enough to hold on to those tricks, once they were discovered. For the most part, what is good enough for the body is good enough for the brain.

OUR PLACE IN THE UNIVERSE

It is popular in some quarters to claim that the human brain is largely unstructured at birth; it is tempting to believe that our minds float free of our genomes. But such beliefs are completely at odds with everything that scientists have learned in molecular biology over the past decade. Rather than leaving everything to chance or the vicissitudes of experience, nature has taken everything it has developed for growing the body and put it toward the problem of growing the brain. From cell division to cell differentiation, every process that is used in the development of the body is also used in the development of the brain. Genes do for the brain the same things as they do for the rest of the body: They guide the fates of cells by guiding the production of proteins within those cells. The one thing that is truly special about the development of the brain—the physical basis of the mind—is its "wiring," the critical connections between neurons, but even there, as we will see in the next chapter, genes play a critical role.

This idea that the brain might be assembled in much the same way as the rest of the body—on the basis of the action of thousands of autonomous but interacting genes (shaped by natural selection)—is anathema to our deeply held feeling that our minds are special, somehow separate from the material world. Yet at the same time, it is a continuation, perhaps the culmination, of a long trend, a growing-up for the human species that for too long has overestimated its own centrality in the universe. Copernicus showed us that our planet is not the center of the universe. William Harvey showed that our heart is a mechanical pump. John Dalton and the nineteenth-century chemists showed that our bodies are, like all other matter, made up of atoms. Watson and Crick showed us how genes emerged from chains of carbon, hydrogen, oxygen, nitrogen, and phosphorus. In the 1990s, the Decade of the Brain, cognitive neuroscientists showed that our minds are the product of our brains. Early returns from this century are showing that the mechanisms that build our brains are just a special case of the mechanisms that build the rest of our bodies. The initial structure of the mind, like the initial structure of the rest of the body, is a product of our genes.

Human brains, presumably, are more complex than those of any other species, but there's little reason to think that the process of their development is wholly or even significantly different from the process by which the brains of other animals develop. As we will see in Chapter 7, the vast majority of the components of the human brain are related to the components of other brains and arise in similar ways. From the perspective of the toolkit of developmental biology, brains are just one more arrangement of molecules. If we accept that our minds are the products of our brains, we must accept that the basic processes by which our minds are built are of a piece with those that build the brains and mental systems of other organisms.

Although some might see the idea that we are just a bunch of molecules, grown in all the usual ways, as a bleak renunciation of all that is special about humanity, to me it is an exciting modern take on an old idea, that there is a bond that unifies all living things. Saint Francis is said to have "called all creatures, no matter how small, by the name of brother and sister, because he knew they had the same source as himself."[64] Where the ancients might have had to point to the supernatural, we can now point to the physical. Through advances in molecular biology and neuroscience, we can now understand better than ever just how deeply we share our heritage—physical and mental—with all the creatures with whom we share our planet.

6

WIRING THE MIND

It's what you learn after you know it all that counts.

—Earl Weaver

WHEN I WAS growing up, Radio Shack sold do-it-yourself electronics kits that were a bit like Legos for the electric set. A kit would have fifty or a hundred components—transistors, resistors, capacitors, diodes, switches, dials, speakers, battery connectors, and so on—built into a big board. Each component was soldered to two or three springs. To use a component, you would bend the associated spring and stick a wire inside; the spring would snap back and hold the wire tight. By connecting the wires in the right way, you could build whatever you wanted—a radio, a flashing light, a siren, maybe even (if you had the biggest kit) a very simple computer. With the same basic components, you could build a lot of different kinds of circuits. What really mattered was how you put them together.

The same is true for the brain. A brain with neurons that weren't connected would be like an electronics kit with no wires—pretty useless. And as with the electronics kit, it wouldn't do much good if the wires were placed at random. The components, in both cases, are important, but they are useless without the connections between them. Indeed, the wiring between neurons is arguably the single thing that makes the brain most special. For it is that wiring that allows the brain to compute and analyze, reason and perceive. The essence of being an intelligent being is the ability to gather information from the world and use that information to sensibly inform action. To do that, an organism's nervous system must transmit information from the senses to higher-level command centers that make choices, and then translate those choices into specific instructions that must be conveyed to the muscles. The billions of neurons in your brain have trillions of connections between them, and what your brain does is largely a function of how those connections are set up. Alter them, and you alter the mind. In the laboratory, mutant flies and mice with aberrant brain wiring have trouble with everything from motor control (one mutant mouse is named *reeler* for its almost drunken gait) to vision. And in humans, faulty brain wiring contributes to disorders from schizophrenia to autism.[1]

The importance of brain wiring to the mind was perhaps most dramatically illustrated in the 1960s when neuroscientists Roger Sperry and Michael Gazzaniga examined the mental function of "split-brain" patients—epileptic patients who as part of their treatment had had the band of 200 or so million connections that run between the left and right hemispheres of the brain—known as the *corpus callosum*—surgically severed. Although these people seemed normal in everyday interactions—they could talk, read, and recognize people and objects—their minds were in fact radically altered. Even though they had little difficulty naming everyday objects that they saw in the right half of their visual field (which connects to the left, more linguistic hemisphere), they could not identify the same objects if they saw them only in the left half of their visual field (which connects to the right, less linguistic hemisphere). In an ordinary person, the two halves of the brain communicate with each other: If the right half of the brain sees a spoon, it passes a message across the corpus callosum to the left hemisphere, and the verbal left hemisphere can name that spoon. Without the connections that run through the callosum, the right hemisphere can't pass its message to the left, and the split-brain patients are unable to voice their observations—even though other nonlinguistic experiments make it clear that the right hemisphere has no trouble perceiving the spoon.[2] In Sperry's words, it was as if the split-brain patients had "two separate realms of conscious awareness; two sensing, perceiving, thinking and remembering systems."[3]

What's true at a macro-level is also true at a micro-level: No individual neural circuit works properly unless it is wired correctly. A mouse can only walk in an alternating left-right-left-right gait if certain neurons in the central midline of its spinal cord are wired up correctly;[4] a male worm can only turn around to find a female if its sensory "ray" neurons are properly wired to tail-flipping motor neurons.[5] Any creature is only as good as its wiring.

What tells the brain how to put its circuits together? Radio Shack kits came with wiring diagrams known as schematics, but human brains don't come with this kind of instruction. If there is no blueprint for building your neurons, surely there is no blueprint for establishing the connections between them. Even in the *C. elegans* worm—where there are just 302 neurons, and only about 7,600 connections among them,[6] the process by which brain wiring is established is a gradual one, directed, like any other aspect of biology, not by diagram but by algorithm, by the action of individually empowered entities, in this case the *axons* (outputs) and *dendrites* (inputs) that extend forth from the neurons.

NEURAL NAVIGATORS/ONE IF BY LAND

How do the brain's axons and dendrites know where to go? Do they have directions that say things like "Head straight for spinal cord, and do not pass Go"? Or do they rely on absolute metrical instructions such as "Keep going the way you are headed for 12 millimeters, and then take your first left"? Much of what goes on is decided by special, wiggly, almost hand-like protuberances at the end of each axon known as *growth cones*.

Growth cones (and the axonal wiring they trail behind them) are like little animals that swerve back and forth, maneuvering around obstacles, extending and retracting little feelers known as *filopodia* (the "fingers" of a growth cone) as the growth cone hunts around in

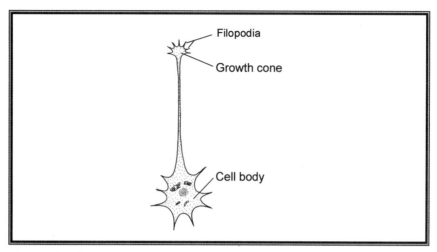

Figure 6.1   Growth cone
Illustration by Tim Fedak

search of its destination. Rather than simply being launched like projectiles that blindly and helplessly follow whatever route they first set out on, growth cones constantly compensate and adjust, taking in new information as they find their way to their targets.[7]

As a whole host of experiments has shown, growth cones can cope even with radical disruptions to the usual situation. In one early experiment, embryologist Emerson Hibbard carved a bit of hindbrain tissue out of a salamander, rotated the tissue, and reimplanted it. The axons of the giant Mauthner neurons (which usually head toward the spinal cord) briefly moved in exactly the opposite direction (toward the forebrain) but somehow detected their error and quickly reversed course, finding their usual destination using an entirely novel route.[8] More recently, William Harris of Cambridge University tested what would happen if primordial eyes were implanted in unusual locations, forcing the optic nerve to enter the brain in front of or behind its usual spot. Neither rain nor sleet deterred the retinal axons: No matter what Harris did, they managed to find their way to their usual targets.[9] Thus, growth cones do not just head in a particular direction and hope for the best. They "know" what they are looking for and can make new plans even if experimentally induced obstacles get in their way. (Amazingly, growth cones can do this more or less on their own. In another classic experiment, Harris severed healthy growth cones from the neural cell bodies to which they were connected. The decapitated growth cones continued on their merry way, twisting and turning toward the usual targets.[10])

In their efforts to find their destinations, growth cones use every trick they can, from short-range cues that require them to get up-close and personal to long-distance cues that act like beacons broadcasting their signals millimeters away,[11] miles and miles in the geography of an axon.

Some of the short-range cues are called *cell adhesion molecules,* and they act like special kinds of glue, of interest only to growth cones that have a matching (often identical) glue. A growth cone with CAD1 (which we can think of as Cell ADhesion glue number 1) will follow only trails that have CAD1, a growth cone with CAD2 will follow only trails that have CAD2, and so forth. (Some trials may be marked off by combinations of different CADs, so that they appeal only to growth cones with the right mix, say CAD1, CAD3, and CAD5.)[12] CADs are great—if you're in the right neighborhood. But they are little use to an axon that is still far from its final destination. CADs are a bit like the solid yellow lines that guide drivers along a highway—not to be ignored, but useful only if the driver has gotten to the right highway in the first place.

Other "long-distance" molecular cues help growth cones find the right neighborhood—the right highway. For example, some of the "radio beacons" that I described in the last chapter can diffuse across long distances, serving as guides to distant growth cones—provided that those growth cones are tuned to the right station. The stations that a growth cone picks up—and whether the growth cone finds a particular signal attractive or repellent—is a function of the receptors it has on its surface.

Consider, for example, the axons that stem from the ventral nerve cord (the insect analog of the spinal cord) in a growing fruit fly. Short-range "highway markers" help large sets of growing axons to stick together into bundles known as *fascicles*—but that only keeps axons from swerving off the road; it doesn't tell them whether they are on the right highway. The choice of highways depends on a family of receptors known as the *"Robo"* family: *Robo, Robo2,* and *Robo3* (no relation to Paul Verhoeven's dark science-fiction masterpiece *Robocop*).[13] Which highway an axon follows depends on which of the *Robo* receptors are expressed on its growth cones (itself a matter of gene expression, since each receptor is ultimately the product of a protein template THEN). Axons that are destined for the inner highway have *Robo* on their growth cones but lack *Robo2* and *Robo3*. Axons destined for the middle highway have *Robo* and *Robo3,* but no *Robo2*. Axons destined for the outer highway have all three. The same code holds even when investigators have altered which *Robo*s are on a given growth cone. A growth cone that is fooled into having all three, for example, will head to the outer highway, even if under ordinary circumstances it would have been destined for the inner highway.[14] Humans have *Robo*s, too, and it is likely that they play a similar role in establishing the architecture of the nervous system of human infants.

Other studies have revealed different codes for guiding different sets of neurons. For example, in a tour-de-force study of the nervous systems of embryonic mice, Samuel Pfaff and his colleagues at the Salk Institute in San Diego took a group of thoracic (chest) motor neurons that normally extend their axons into several different places, such as axial muscles (midline muscles that play a role in posture), intercostal muscles (the muscles between the ribs), and sympathetic neurons (which, among other things, participate in the fast energy-mobilization for fight-or-flight responses), and by changing their genetic labels persuaded virtually the entire group of thoracic neurons to abandon their usual targets in favor of the axial muscles.[15] (The few exceptions were a tiny number that apparently couldn't fit into the newly crowded axial destinations and had to find other targets.)

If axons are the "output" wires of a cell, dendrites are the "input" wires, and synapses are the places where the two sorts of wires meet. Dendrites have been less well-studied, probably because it was once thought that they merely waited passively for their axons-in-shining-armor to come riding in. But Liqun Luo, a developmental neuroscientist at Stanford University, has shown that dendrites are as active in the process of synaptic matchmaking as their more prominent axonal neighbors.[16] For example, in the olfactory system of a growing fruit fly, there are two major classes of excitatory neurons, known as "adPNs" and "lPNs," each of which sends its dendrites to distinct locations to receive input from specific olfactory sensory neurons. Luo and his colleagues have shown that those choices are governed largely by just two genes, known as *acj6* and *drifter*. Under ordinary circumstances, adPNs express (that is, switch on) *acj6*, while lPNs express *drifter*. By altering where they were expressed (for example, by knocking them out or switching them around), Luo's team altered the patterns of dendritic connections, much as Pfaff's group had altered the targets of axons.

What this all boils down to, from the perspective of psychology, is an astonishingly powerful system for wiring the mind. Instead of vaguely telling axons and dendrites to connect at random to anything else in sight, which would leave all of the burden of mind development to experience, nature supplies the brain's wires—axons and dendrites—with elaborate tools for finding their way on their own. Rather than waiting for experience, brains can use the complex menagerie of genes and proteins to create a rich, intricate starting point for the brain and mind.

When a termite builds an elaborate castle, it does so not because it has taken a master class in castle-building but presumably because nature has evolved its brain and genome with sufficient precision to specify the detail that is required by a termite's nervous system. And as far as anyone can tell, that level of precision is even greater in humans. The precision with which an axon can find its way to its destination depends on its ability to sniff out just the right kinds of signals; the more distinct the signals, the more precision there is. At least half a dozen major families of molecules play roles in axon guidance, and, as we will see in Chapter 8, one of the major findings of the Human Genome Project has been the extent to which the set of such signals has expanded in the evolution of vertebrates (amphibians, fish, reptiles, birds, and mammals), and perhaps especially in humans. If our minds are more complex than the minds of other animals, it is in part because we have more ways of using genes to precisely shape the wiring of our brains.

AXONAL PREDESTINY?

Although the business of axon guidance is often conducted with laser precision, it is not at all preordained, any more than are the exact locations of the millions of tiny arteries, veins, and capillaries in the circulatory system. The intricacy of the final pattern of neural connections comes not from a blueprint or wiring diagram but from the precision of the underlying genetic toolkit, the signals and receptors that guide individual growth cones. Whether we speak of the nervous system or the circulatory system, what is given in the genome is, as we have seen, more method than picture, more recipe than blueprint. Rather than specifying every twist and turn in advance, the genome exploits a far more flexible strategy: It provides recipes for building a particular type of structure whenever it may be needed, without committing in advance to how many of those structures will actually be required.

Your body grows enough skin to cover your bones, and if, for example, a genetic mutation should lead a growing embryo to grow an extra digit, say a sixth finger, the body can supply that finger with the basic utilities—the plumbing and wiring of the blood vessels and neurons—without requiring the genome to specify every detail in advance. Instead, the genome provides general techniques that get invoked as needed, providing an automatic measure of flexibility.[17]

Laboratory studies make that flexibility even more apparent. Consider the "barrel fields" we saw in the last chapter. In general, each rat whisker is represented in the brain by a particular cortical structure known as a barrel field. One could imagine the genome specifying the whiskers and the barrel fields separately; instead, the brain seems to build the barrel fields in response to the nerves that come in from the whiskers. In every strain of rat that has been genetically engineered to have extra whiskers, extra barrel fields develop.[18] The system for assembling the brain can thus readily adapt itself to remarkable changes in the sensory system.[19] (The three-eyed frogs I described in Chapter 3 are an even more dramatic example of this point, since the ocular dominance columns they grow under the laboratory conditions presumably never occur in nature, yet some aspect of the toolkit for neural growth allows the frogs to accommodate to their new sensory apparatus.)

The genome doesn't even have to know in advance exactly how much brain tissue there is. Rather than being tied to absolute markers of position ("Go twelve cells dorsally and hang your first left"), many cues to cortical layout are specified in relative terms ("When you get to where you have only two-thirds as much *Emx* protein, start building your auditory fields"). Like a recipe that can be halved or doubled, the basic instructions for building a mammalian brain seem to be equally at home building a large brain or a small one; as at home in a hamster as a human. The instructions can even, as University of California at Davis neuroscientist Leah Krubitzer recently showed, be modified on the fly. In an experiment in which Krubitzer's team removed half the primordial cortical sheet in growing short-tailed opossums *(Monodelphis domestica),* the resulting embryos grew brains that appeared to have all the normal cortical areas, ordered in the usual way (visual cortex behind auditory cortex behind somatosensory cortex), but compressed into half the usual space.[20] Using the same basic plan but adjusting the dials, neural organization can be fine-tuned over time as a species adapts to its particular niche. The hippocampus, presumed to be involved in spatial memory, is enlarged in birds that remember the locations of large numbers of seeds and in male voles that traverse large territories looking for mates;[21] the olfactory bulb shrinks in carnivores such as otters and whales that spend a large part of their lives underwater, where a sense of smell is less important.[22] By tying development to relative rather than absolute cues, evolution has provided us with a handbook of self-assembly both stunningly flexible and sublimely powerful; flexibility emerging as an intrinsic part of the design.

REWIRING THE MIND

Now, here's the rub. Every genetic process is triggered by some sort of signal. *From the perspective of a given cell, it doesn't matter where that signal comes from.* The signal that launches the adjust-your-synapse cascade, for example, may come from within, or it may come from without. The same genes that are used to adjust synapses based on internal instruction can be reused by external instruction.[23]

The consequences of this subtle fact are gigantic. It would be no great exaggeration to say that it is a—maybe even *the*—key secret to intelligent life on earth. The reason that animals can learn is that they can alter their nervous systems on the basis of external experience. And the reason that they can do that is that *experience itself can modify the expression of genes.*

The role of genes is not just to create the brain and body of a newborn, but to create an organism that is flexible enough to deal with an ever-changing world. Genes play an important role throughout life, not just until the moment of birth, and one of the most important ways in which they participate throughout life is by making learning possible.

To take one example, even a brief exposure to light in a newborn kitten, rat, or monkey can launch a complex cascade of gene expression.[24] The light activates photoreceptors—which send signals—which trigger a pathway—which leads to the expression of neural growth factors and a set of genes known as "immediate early genes" or "early response genes"—each of which, in turn, triggers the expression of many more genes. One study of cichlid fish suggests that a change in social status (from submissive to dominant) is tied to changes in the expression levels of at least fifty-nine different genes[25]—a phenomenon not entirely unrelated to the testosterone rush that Joe-six-pack gets when the home team wins.

Neural activity can modulate everything from the distribution of receptors and axon guidance molecules to gene expression,[26] in turn guiding everything from cell migration to the tendencies of axons and dendrites to branch.[27] For example, in a neural version of the old saying "use it or lose it," neurons in a dish are more likely to thrive and to divide if they receive repeated electrical stimulation.[28] In living organisms, the brains of rats and mice that are raised in complex, toy-filled environments have, in comparison to rodents raised in ordinary, drab cages, thicker cortical tissue, more intricately branched dendrites, and more synapses per neuron.[29] Putting a rat in an enriched environment for just three hours leads to increased expression of at least sixty different genes, genes that increase DNA replication, guide the growth of synapses, and reduce cell death.[30] Just as exercise causes new blood vessels to sprout,[31] learning may co-opt the molecular cascades that cause the brain to produce new synapses.[32] Learning, whether in a rat or a human, is a process by which experience modifies the brain by modifying the expression of genes.

Not every gene, nor every brain connection, can be modified by experience. Each species has different ways of connecting experience to gene expression, and these different links make possible different kinds of learning. The point here is not that genes allow us to learn just *anything,* but that whatever we do learn is made possible, in one way or another, by specific genetic mechanisms. Whether a particular species can learn a song or a sentence depends on the IFS and THENS that make up its genome.

THE MIND'S BLACKBOARD

Regardless of which species we talk about, or which aspect of mental life we investigate, the ability to learn starts with the ability to remember. An organism can learn from experience only if it can rewire its nervous system in a lasting way; there can be no learning without memory. Most research on the biology of memory has focused on something I'll call "synaptic strengthening." Synapses, the connections between one neuron and the next, are thought to vary in strength, with strong connections between neurons that are in some way closely tied together. Let us suppose that a simple organism has one neuron for recognizing a special sound, call it the "bell neuron," and another for triggering the complex set of cells involved in eating, call it the "munch neuron." The bell neuron would fire whenever the simple creature heard the bell, the munch neuron whenever the creature began to eat. If the animal was consistently fed right after the bell rang, one might expect that, over time, the connection—the synapse—between the bell neuron and the munch neuron would get stronger, making the creature more likely to want to munch when-ever it heard the bell. Indeed, Pavlov's famous experiments with dogs in the early twentieth century suggested exactly this.

And nature does seem to have a process that explains his findings. It is now known as LTP, which stands for "long-term potentiation." The idea is that certain kinds of learning might depend on "potentiating"—strengthening—the synaptic connections between neurons.[33] This process of strengthening a synapse is long and complex—more than a hundred different molecules may be involved,[34] and there are at least fifteen distinct steps in the process, but they can be roughly divided into five basic stages.[35] First, the brain notices that something interest-ing has happened and some neuron "fires," releasing neurotransmitters on the "transmitting" side of the synapse. Next, the neurotransmitters that are released on the transmitting side bind to appropriate receptors on the receiving side of that synapse. Those receptors then allow charged atoms through. Once inside the cell, those charged atoms launch a biochemical cascade that ultimately switches on a set of early-response genes. Those early-response genes then ultimately launch a *second* round of gene expression, which in some way (still under inves-tigation) physically strengthens the synapse, quite likely by using many of the same genes and proteins (such as cell adhesion molecules) that direct initial synapse formation endogenously, prior to experience.

Each stage in the process of memory formation has a genetic component.[36] The receptors that respond to neurotransmitters, for example, are proteins, and interfering with them interferes with memory. Genetically engineered mice that have been designed to lack specific kinds of receptors have trouble with specific kinds of learning,[37] and the same holds for mutants that lack other proteins that are important in the process of memory formation, such as CaM Kinase II (a calcium-activated enzyme for energy transfer and signaling).[38] Indeed, the whole process of protein building is essen-tial for long-term memory formation, and interfering with that pro-cess can lead to amnesia for specific events,[39] and even preventing songbirds from learning new songs.[40]

Through judicious genetic tinkering, memory can be, at least to some extent, improved. A 1999 study showed that mutant mice that have extra NMDA receptors—special receiving-end "coincidence" receptors that appear to excel at noting when two things happen at the same time—have *better* memories than normal mice.[41] The newspaper headlines—"Scientist Creates Smarter Mouse"—were, as usual, a bit of an exaggeration: There were no permanent gains, and no evidence that the mice really were smarter. Mice with extra NMDA receptors did better on some short-term memory measures, outperforming controls on tests that required them to recognize objects. But the gains were fleeting, lasting for a few days, and disappearing by the time the mice were tested a week later, suggesting that the extra receptors help with some intermediate process rather than with the ultimate consolidation into permanent memory.[42] Mind you, even if the results were stronger, I wouldn't recommend that you try injecting NMDA receptors at home. It is a good bet that there is a reason nature hasn't endowed us with massive quantities of them, at least one of which became clear when later studies revealed that the mutants were also more sensitive to inflammatory pain.[43] You probably don't want to remember *everything* better.

Roughly the same sets of molecules seem to play more or less the same roles in just about every organism that's been studied. As far we can tell, whether a chick is recording the appearance of its mother or a songbird is learning a new song, the mechanisms of information storage appear to be the same. Immediate early genes and NMDA receptors, for example, seem to contribute to the imprinting of a chick, the aversion to a taste that has induced nausea in a rat, and the song-learning ability of a sparrow.[44] As psychologist Randy Gallistel has put it, "Information is information": "An important principle in modern computing and communication is that different kinds of information are equivalent and interchangeable when it comes to storage and conveyance; a mechanism suited to store or convey one kind of information is equally well suited to store or convey any other kind."[45] If memory is like a blackboard, it seems likely that most mental processes may use the same kind of chalk.

We know something about what that chalk might be, but there is plenty left to be discovered. We know little about the mechanisms by which memories are retrieved, and even less about the "codes" the brain uses to store its memories; it is as if we understood the process by which chalk is applied to blackboards, but nothing of writing or how it is read. Even when it comes to the synapse-strengthening process that I described earlier, which is the best-understood of the neural processes related to memory and learning, there is much that has not yet been resolved.[46] It is not yet clear whether the process of synaptic strengthening truly plays a role in long-term memory, or whether it plays a temporary role only in an intermediate consolidation from short-term memory to long-term memory. And although a great deal of evidence suggests that the genes involved in LTP are *necessary* for memory, there is as yet no demonstration that they are *sufficient* for memory. Other genes may well be involved, especially in the formation of permanent memory, which some researchers have suggested might rely not only on changes in synapses but on other mechanisms, such as changes in DNA itself.[47]

Though there may be just one kind of chalk, there is surely more than one blackboard. Neural substrates for memory are found not just in one particular location in the brain, but spread throughout, with different circuits supporting different kinds of memory. Memory systems can be found not only in the hippocampus (which has some role in spatial memory) but also in the cortex,[48] the amygdala,[49] and in a variety of visual and motor areas.[50] Although the same general cascades of biochemical processes—from the binding of receptors for neurotransmitters to the activity of early-response genes to the modification of synapses—take place in each memory system, each memory system has a different function. Memory in the hippocampus has to do with spatial locations, for example, whereas memory in the amygdala has to do with emotional events. Selectively lesioning the hippocampus in a rat selectively impairs the rat's spatial memory.[51] Impairing its amygdala impairs its emotional memory.[52]

Disruptions to memory systems can have very different effects depending on which memory system is disrupted. By selectively disrupting CaM Kinase II (that energy-transfer/signaling enzyme mentioned above) in two different brain locations, Nobel laureate Eric Kandel created two different types of mutant mice: "hippocampal mutants" with impaired spatial memory and "amygdala mutants" with impaired emotional memory.[53] (The term "such-and-such mutant" is laboratory shorthand for an animal, here a mouse, that has been genetically engineered to have a disruption in a particular region of the brain. A hippocampal mutant is an animal that has been genetically engineered to have a disruption in a particular gene normally expressed in the hippocampus, an amygdala mutant one that has been engineered for a disruption in the amygdala.)

The hippocampal mutants had little difficulty learning to fear a tone (or novel environment) that was paired with a foot shock, yet they couldn't find their way out of a circular maze that always had the same exit, even after forty days of practice. The amygdala mutants easily mastered the circular maze, but they never learned to fear the tone that warned of the shock. Imaging studies in humans show similar specialization. Although each memory system appears to use more or less the same set of molecular mechanisms, different cognitive systems store their memories in different places. Same chalk, different blackboards.

BEYOND MEMORY

Studies of memory go part of the way—but only part of the way— toward helping us understand specialized learning mechanisms. It is likely that each specialized learning mechanism relies on its own specialized memory store. But another part of learning comes in deciding which information to store in the first place. The next step in our scientific understanding will be to discover how a newborn chick "knows" to prefer objects with necks and shoulders, or how the newborn indigo bunting "knows" to look for the rotation of the stars.

Though little is known about these neural substrates that allow organisms to know what to look for, there already are some intriguing hints. We know that the newborn chick looking for its mom actually relies on at least two different neural systems, one for orienting toward stimuli that are good candidates for being mom, and another for taking whatever it can get, for storing a memory of anything that the chick might be exposed to. The first system will choose an adult chick (or even a stuffed duck) over a box as its go-to caregiver, but if there's nothing else around, the second system will lead the chick to settle for the box.[54] In the next decade, we'll likely find out much more about how such orienting systems work and how genes contribute to their development.

Orienting toward the right bit of information is only the first step. A taste aversion system must not just identify and categorize tastes, it must somehow supply that information to the cognitive systems that control an animal's dietary preferences. The indigo bunting's celestial system must take what it learns about the stars and feed it into a navigation system that sets the bird's heading. A swamp sparrow's song-learning system must identify the right songs to learn and somehow decompose those songs into notes and phrases before it can use that information to tune its own song.

Song-learning systems are especially interesting because they are so similar in abstract design to our own linguistic system. Learning a song appears to require separate systems ("modules") in the songbird for detecting which songs are from its own species, for parsing those songs into notes and phrases, for recording the components of the songs in memory, and for turning those stored representations into vocal gestures. One part of the bird's front forebrain, known as "LMAN," seems to contribute to learning—but not performing—songs. If that part of the brain is removed in a young zebra finch, its song will be frozen in a premature state; yet removing it in adults does not prevent them from singing songs they already knew.[55] Another important part of the system is the pathway (that is, the set of neural wires) that runs from the avian counterpart to human language areas to a motor area known as "RA." This pathway plays a role in translating memorized songs into movements by the vocal tract. If the pathway is disrupted, performance on all songs—old and new—is radically impaired.[56] As in other systems of learning, genes play a critical role. Early-response genes, for example, are triggered more when a bird hears a song from a fellow member of its own species than when it hears a song of a different species.[57]

Although many of the details remain to be worked out, the general outline of the story seems clear: The bird breaks down the process of learning into several subtasks, each supported by a separate bit of neural circuitry. Learning itself is likely to be a process of using experience to tune the modules and the connections between them—mediated, always, by genes. It would not be outlandish to expect a roughly similar story in the human being: sets of distinctly specialized neural systems, each participating in different parts of the language-learning problem, tuned by experience that is mediated by the action of genes.

The genetic side of the process remains speculative, in part because of the technical limitations involved in conducting experiments with birds (there is not yet an easy way to alter their genomes), and the ethical ones with humans (not even a mad scientist would dare to study the effects of knocking out synaptic strengthening in Broca's area). But one organism that the geneticists do know a lot about is the lowly *C. elegans* roundworm, and what we know about it fits well with the overall picture I have been sketching. Even in the roundworm, learning is not due to a single, all-purpose mechanism: Worms use different learning mechanisms for different tasks. And in the case of the worm, scientists are making significant progress in understanding the genetic basis of different learning mechanisms. As you might by now guess, the molecular mechanisms for memory are at least partly shared from one type of learning to the next,[58] but each learning mechanism also depends on its unique genes.[59]

For example, neurobiologists Glenn Morrison and Derek van der Kooy at the University of Toronto have found two mutants, dubbed *"lrn1"* and *"lrn2,"* that have trouble with associative learning. Roundworms are naturally drawn to the odor given off by the compound known as diacetyl (a compound that can give beer a butterscotch flavor), but normal worms learn to avoid it when it is paired with an aversive acetic acid solution. *Lrn* mutants cannot do this. Yet they are able to do a different kind of learning (known as habituation) perfectly normally: When they are first exposed to the diacetyl, they track it assiduously, but after fifteen minutes, they learn (even in the absence of acetic acid) that there's no real butterscotch at the end of the rainbow and begin to ignore the misleading diacetyl. The *lrn1* and *lrn2* mutants also *dishabituated* normally. After testing the worms for habituation, the experimenters took the diacetyl away and distracted the worms by putting them in a high-tech salad spinner for sixty seconds. After the wash cycle was over, the mutants recovered their interest in diacetyl, going for it just as much as they did before habituation, and just as much as spin-washed controls.[60]

Association and habituation are among the most basic processes involved in learning, but, as the worm studies make clear, they are not identical, and in fact they depend in part on different genes. A 2003 review written by Columbia University neuroscientist Oliver Hobert reports that there are at least seventeen genes involved in different aspects of worm learning.[61] It is no exaggeration to say that genes are essential to nearly every aspect of memory and the process of learning; without them, learning itself would not exist.

∞

In time, an understanding of learning-related genes may give researchers clues to the causes of learning disabilities. For example, one severe disorder, known as neurofibromatosis, has already been tied to the excessive production of an enzyme that mediates a cascade from receptor to gene expression, perhaps thereby interfering with synaptic strengthening.[62]

Studies of the genes involved in learning may also eventually lead to insights into why the ability to learn certain things diminishes over the life span. The saying "You can't teach an old dog new tricks" has an element of truth to it. Adult dogs really are harder (if not impossible) to train,[63] and adult human beings aren't nearly as good as children at mastering new languages[64] or picking up musical instruments.[65] Similar results have been shown under laboratory conditions. Stanford biologist Eric Knudsen, for example, has shown that barn owls are better able to recalibrate their ears to their eyes early in life than later in life.[66] Since a barn owl captures much of its prey at night, it relies heavily on a precise sense of hearing, which in turn is tuned by visual feedback. When Knudsen put prismatic glasses on the eyes of young barn owls, they quickly retuned the mapping between eye and ear, accommodating themselves to a strange new visual world; adults, by contrast, showed much more limited abilities to perform such retuning.

Adults are not a lost cause: You are learning (and rewiring your brain) even as you read this book. Dozens of studies over the past few years have shown that the brains of adult animals have more "plasticity" than was once thought, as in the study with monkeys mentioned in Chapter 3, in which the parts of the cortex that were originally devoted to an experimentally amputated finger eventually came to respond instead to the neighboring (intact) fingers.[67] But the ability to learn does indeed vary over time, diminishing more sharply in some domains than others. From an evolutionary perspective, this makes sense; an animal should be able to learn new things about its environment throughout life, but once its body stops growing, it shouldn't need to recalibrate its hand-eye coordination on a daily basis. If learning is costly—all that readjustment takes energy, and might break a system that is already working—it might be advantageous to shut it down after a certain point.

A major push is under way to figure out the molecular basis of those "critical" or "sensitive" periods, to figure out how the brain changes as certain learning abilities come and go. In some, if not all, of those mammals that have the alternating stripes in the visual cortex known as ocular dominance columns, those columns can be adjusted early in development, but not in adulthood.[68] A juvenile monkey that has one eye covered for an extended period of time can gradually readjust its brain wiring to favor the open eye; an adult monkey cannot adjust its wiring. At the end of a critical period, a set of sticky sugar-protein hybrids known as *proteoglycans* condenses into a tight net around the dendrites and cell bodies of some of the relevant neurons, and in so doing those proteoglycans appear to impede axons that would otherwise be wriggling around as part of the process of re-adjusting the ocular dominance columns; no wriggling, no learning. In a 2002 study with rats, Italian neuroscientist Tommaso Pizzorusso and his colleagues dissolved the excess proteoglycans with an antipro-teoglycan enzyme known as "chABC," and in so doing managed to re-open the critical period. After the chABC treatment, even adult rats could recalibrate their ocular dominance columns.[69] ChABC proba-bly won't help us learn second languages anytime soon, but its anti-proteoglycan function may have important medical implications in the not-too-distant future. Another 2002 study, also with rats, showed that chABC can also promote functional recovery after spinal cord injury.[70]

AUTODIDACT

Remarkably, many of the brain's mechanisms serve double duty. The mechanisms that allow the brain to rewire itself on the basis of expe-rience actually get exploited even before we have contact with the outside world because they also respond to *internally generated* expe-rience. Monkeys, for example, take a first step toward stereo vision by setting up their ocular dominance columns in a darkened womb. They probably do this both by using internal molecular cues that act in an experience-independent way and by spontaneously generating their own "experience."

In sensory systems ranging from vision to audition and somatosen-sory sensation, and in organisms from turtles to mice,[71] scientists have found that embryonic vertebrate brains spontaneously generate neural activity even before their senses are hooked up to the outside world[72] and that this self-generated activity allows embryonic brains to refine their own wiring.[73] Scientists have discovered traveling "waves"—pat-terns of electrical activity (apparently modulated by a signaling mole-cule known as "cyclic-AMP")—that sweep across the retina[74] every minute or so at the rate of about 100 micrometers per second,[75] and they have uncovered similar prenatal oscillations that travel through the cochlea, spinal cord, hippocampus, and cortex.[76]

Waves work their magic by exploiting some of the same mechanisms—such as cascades driven by coincidence-detecting NMDA receptors—that allow the brain to learn from the external world.[77] Like the multicolored test patterns broadcast on late-night television, waves provide a known signal that can be used to calibrate the machinery. In the Society of Motion Picture and Television Engineers color test pattern, you adjust your set so that the rightmost stripe looks blue, the stripe to its left looks red, and so forth. In the case of cyclic-AMP-driven waves, your brain does the work of rewiring itself, ensuring that neurons that are near one another represent perceptual input from locations that are close together. Experiments with strobe lights,[78] artificially electrical stimulation,[79] and chemicals that change cyclic-AMP levels[80] all show that disrupted waves lead to disrupted wiring.

PUTTING IT ALL TOGETHER

As we've seen, from the perspective of a neuron, it doesn't matter whether a signal comes from outside or inside. Information is information, and evolution has wired our embryonic minds up to use it all in the same way. Electrical and chemical activity can mediate many of the same processes that make growth possible in the first place. Whether spontaneously generated on the inside, or driven by experience with the world, electrical activity works together with genes to help shape the fates of neurons and the connections between them.

Learning proceeds not by overriding the genes (which would be a case of nurture emerging triumphant, with genes rendered mute) but by repurposing them, by adapting ancient techniques of development for modern needs of on-line flexibility. In this way, genes are as important to learning as they are to "innate" development. If brain wiring is a special case of the toolkit of embryology, learning is a special case of brain wiring, as much a product of the genes as any other aspect of biology.

At the end of *Portnoy's Complaint,* in the very last sentence of Philip Roth's famous 1969 novel, after a recap of Portnoy's life, Portnoy's psychoanalyst finally speaks his first line, "So. Now vee may perhaps to begin. Yes?" In the last three chapters, we have seen the steps by which the brain first organizes itself. But that first organization, driven by genes, is just the beginning. Genes continue to contribute to everything that follows. Once the brain's first systems come on-line, a new stage of development begins. Genes, at last, take on the outside world as an equal partner, and the two begin to parlay their combined forces in the pursuit of a mind entirely new.

An ancient Chinese proverb says, "Give a man a fish and he will eat for a day. Teach a man to fish and he will eat for the rest of his life." Nature has followed that sentiment by devoting a healthy chunk of the genome to mechanisms for making the brain—and by extension, the organism—flexible enough to fend for itself.

But where did these genes come from?

7

THE EVOLUTION
OF MENTAL GENES

A zoologist from Outer Space would immediately classify us as just a
third species of chimpanzee, along with the pygmy chimp of Zaire and
the common chimp of the rest of tropical Africa. Molecular genetic
studies . . . have shown that we . . . share over 98 percent of our genetic program with the other two chimps.

—Jared Diamond

IF YOU ARE by now convinced that genes play an important role in
shaping the mind, you might be curious about what shapes the genes.
Where do the genes that participate in the building and maintenance
of the brain come from? Like the genes that build the body, the genes
that build the brain are a product of evolution. Complex organs like
the eye and the brain developed not overnight but over the course of
millions or billions of years, gene by gene, protein by protein. With
modern biological techniques it is possible to begin to reconstruct
that history, to figure out when the various components of the brain
first appeared on the scene. My goal in this chapter is not to consider
what the brain evolved *for*—an ever-controversial topic that is outside
the scope of this book—but to explain how (and when) the genes that
help to build the brain evolved.

All evolution arises in one way or another from some kind of change in the genetic code. The most familiar kind of genetic change is the simple mutation, an A changed to a C, a T to a G. As we have seen, such mutations can lead to disorders, but they can also lead to useful evolutionary innovation. On occasion, a mutation—which might result from radiation, toxic chemicals, viruses, or errors in the process of DNA replication—turns out to be a good thing, something that helps its bearer have a better chance of thriving and reproducing. A particularly valuable mutation may gradually spread through the population; such is the source of much evolutionary change.

Sunspots, viruses, and plain old copying errors can also lead to another kind of change: They can cause nucleotides to be inserted (for example, AG becomes ACG), deleted (ACG becomes AG), or inverted (ACG becomes GCA), and the same sort of thing can happen with larger chunks of chromosomes. Perhaps less familiar is a mechanism known as *duplication*. Errors that occur during the process in which genetic information is copied or prepared for transmission from parent to child can inadvertently lead to the duplication of an entire gene, an entire chromosome, or even an entire genome, leading the child to have two copies where a parent had one.

At first glance, you might well wonder why such an event—a duplication—could matter. It's easier to imagine how insertions, deletions, inversions, and even substitutions might matter, since each of those processes can lead to an immediate change in the corresponding amino acid. A change of an AGC to an AGG, for instance, leads a molecule of the amino acid serine to be replaced with a molecule of a different amino acid, arginine. Since serine and arginine have different molecular structures, it is not hard to imagine how a change from a C to a G might lead to an important change in protein structure, maybe good, maybe bad, but in any case something that could influence an organism's chance of producing viable offspring. But why should it matter if an organism should suddenly have an extra copy of a gene? A cookbook with two identical copies of the recipe for tofu lasagna is no better than a cookbook with single copy.

One reason that duplications matter is that an extra copy of a gene can mean an extra chance to make a particular protein. At the molecular level, genes are a bit like lottery tickets; rather than giving a guarantee of something, they give only a probability. A gene provides a key to synthesizing a particular protein only if the right keys that open the regulatory locks fall into place in just the right way. The constant pushing and jostling of molecules means that the keys don't always at every moment fall perfectly into those locks; an extra copy of the gene may increase the chance that the corresponding protein gets made. An extra copy can also mean that twice as much of the protein is made, yielding, say, a more rigid cell wall or an increased gradient of regulatory proteins, which might change the relative proportions of two bones. Making extra copies isn't always advantageous—there might be a benefit in the more flexible cell wall, but there might be a disadvantage to fiddling around with the ratio between femur and tibia. Turning to the brain and mind, several kinds of mental retardation, such as Down syndrome (Trisomy 21) and Patau syndrome (Trisomy 13), appear to be caused by superfluous copies of genes.[1]

But there is an even more important reason why duplication may have had a large impact on evolution—it provides what Richard Dawkins described as the blind watchmaker—evolution—with a way around the old adage of "if it ain't broke, don't fix it." If one copy of the gene—perhaps already optimized to a particular function—remains stable, the second may vary without loss of the initial function, ultimately giving rise to new function.

Though none of this is done by forethought, the consequences can be profound. Our ability to see color, for example, appears to have depended on two such duplications. Some of our earliest vertebrate ancestors had only a single type of photoreceptor pigment, one that responded most strongly when it was most illuminated, and less strongly when it was less illuminated. Around 400 million years ago, early in the history of vertebrates, and before the modern classes, such as mammals, birds, and amphibians, emerged, the genetic recipe for that photopigment was, by sheer chance, duplicated. When one copy of the randomly duplicated gene diverged (that is, changed slightly through some process of mutation), a new kind of photopigment developed, one that was sensitive to a different part of the light spectrum. With two types, it became possible (given some further machinery to interpret the output of the photoreceptors) to discriminate shorter wavelengths of light (such as blue and purple) from longer ones (such as red and green). About 35 million years ago, one of our primate ancestors branched away from other mammals when there was a second gene duplication, this time of the genetic recipe for the long wavelength ("red") photopigment, leading to a third type of photopigment and what is known as trichromatic vision.[2] (Some versions of color blindness stem from mutations to one or another photopigment; at the opposite extreme, some women have a fourth photopigment, giving them an edge when it comes to recognizing subtle variations such as mauve and chartreuse.[3])

As we shall see, the coordinated processes of duplication and divergence have played an important role in nearly every step in the evolution of the brain. New forms in evolution almost never arise from scratch; they are almost always variations on a preexisting theme.

PIECING TOGETHER THE BRAIN

A major part of what the brain does is to communicate signals from one place to another. It takes information from the senses, analyzes that information, and translates it into commands that get sent back to the muscles. Although in the grand scheme of 3.5 or 4 billion years of life on earth,[4] the brain per se is a relatively recent innovation—perhaps only half a billion years old in a close relative of a pinheaded anchovy called *amphioxus*[5]—many of the brain's components are far older. Organisms as simple as the sponge *Rhabdocalyptus dawsoni* have the rudiments of a recognizable nervous system,[6] and some of the brain's components are even older.

Many single-celled organisms, for example, profit from systems for internal communication. Some bacteria can move toward light or heat, an ability that depends on communication between sensors that take in information from the environment and protein motors that propel the bacteria in the right direction (or at least get them to flip direction when the conditions seem to warrant it). Amazingly, some of the molecules used more than a billion years ago by ancestral bacteria to coordinate information and action remain with us today, in the form of ion channels (those protein gates that open and close so as to control the flow of electrically charged molecules across the borders of a cell).[7] Such channels are found in virtually all living organisms and are major determinants of neuronal function, modulating the sensitivity of individual neurons to factors such as temperature and voltage and playing a role in everything from motion in paramecia to growth in plants and cognition in people.[8] Channels specialized for the flow of potassium probably arose first, but it wasn't long before duplication and divergence led to new classes of channels, each specialized to control the flow of different types of ions (for example, some for calcium, others for sodium or chloride). As R. M. Harris-Warwick put it, "Once one channel gene was made, others could be generated by duplication, allowing diversity to arise in the 'new' copy with no loss of function in the 'old' one."[9]

Further mutation, duplication, and divergence led to receptors, the "receiver" molecules that serve as go-betweens, transforming signals from outside a cell into molecular events inside the cell. These, too, duplicated and diverged early in evolution, creating a variety of receptors, each specialized for receiving a particular kind of signal, such as glutamate, GABA, acetylcholine, or serotonin.[10]

Receptors evolved hand in hand with the signals that they receive. At around the same time that receptors began to evolve, nature developed signaling molecules such as neurotransmitters and neuropeptides. For every signal, there is at least one matched receptor, but different signals control different things. Insulin is a signal that controls the level of blood sugar, adrenaline a signal that prepares the body for action. In some sense, the "meanings" of such signals are arbitrary, in much the way that words are arbitrary. I call a meowing four-legged creature *cat;* Spaniards call it *gato.* There's nothing intrinsic to the structure of insulin that makes it, rather than adrenaline, the signal for controlling blood sugar, and some alien species could easily reverse the role of the two. But the arbitrary conventions adopted by our ancient ancestors are widely shared across the animal world. At a molecular, genetic level, many of the neural signals that we use are nearly a billion years old, and they are found even in bacteria.[11] The human brain may be the best information processor on the planet, but some of the basic signals it uses to process information are almost as old as life itself.

Of course, if you've only got one cell, your communication needs are limited. You don't need to pass messages very far, very quickly, or very precisely. Opening and closing channels are steps in the right direction, but the human brain, with its tens of billions of cells, could never get the job done using channels alone. Just as manmade communication systems have gotten faster, more precise, and more potent over the years, biological systems for communication (both internal and external, although my focus in this section happens to be on communications *within* an organism) have steadily improved over the course of evolution.[12] A smoke signal travels neither fast nor far, and it cannot be aimed toward a particular person. A telephone call placed over fiber-optic cables, in contrast, can travel at the speed of light to a particular person halfway around the world. In a single-celled organism, fast and precise communications may be of a little value, but in a complex, agile organism that has many kinds of cells specialized to many different tasks, fast and precise communications are essential.

As we saw in the last two chapters, electrical impulses are biology's signaling medium of choice. They can travel quickly, and, with the help of those thin, wire-like cables known as axons, they can be directed with precision to particular targets at great distances. When a neuron "fires," it launches a "spike" of electrically charged atoms that travel rapidly down the axon, culminating in the release of the neurotransmitters that allow one neuron to communicate with another.

That biological version of electrical signaling goes back nearly a half a billion years, to jellyfish (or to some closely related ancestor common to us and them), when a growing trend toward cellular specialization led to the development of neurons.[13] Jellyfish nerve cells are far more primitive than ours; their signals travel hundreds of times more slowly than ours do, and their nerve cells have to wait longer than ours do before they can fire again. Moreover, the dreaded jellies don't have anything like a centralized brain; instead, they have a loosely strung collection of neurons that biologists refer to as a mere "nerve net." Primitive, but enough neuron power to control a jellyfish's swimming. The basic plan of our neurons is not so different from theirs. Our nerve cells, like theirs, rely on protein channels that can be modulated—opened and closed—by changes in voltage, and DNA analyses have shown that some of the recipes for building those channels go back at least as far as our common ancestor.[14]

After electrical signaling, the next major step in the evolution of brains like ours was a mix of centralization and bilateralization (a trend toward a left-right split)—organizing principle that apparently began a little over half a billion years ago with a precedent-setting flatworm.[15] Flatworms have muscle cells, sensory cells, and a central set of neurons close to the front end of their body, a division between left and right clusters of neurons connected by a proto–corpus callosum, and three to five sets of nerves—precursors to the spinal cord—that run the length of its body—not so unlike our own system. Although the nervous system of a flatworm is vastly less complex than ours, the resemblance in overall organization is striking,[16] a consequence of the fact that many of the genes guiding the pattern of the brain of a human relate closely to genes involved in the patterning of the nervous system of the worm.[17]

The flatworm's rudimentary division of neural labor—between the central and the peripheral, the left and the right—was the first step toward an avalanche of specialization that has given risen to the complex neural systems of the vertebrates—fish, mammals, birds, amphibians, and reptiles. The vertebrate nervous system radically differs from its predecessors in at least two major ways. First, shortly after vertebrates came on the scene,[18] intercellular communication got a whole lot better, with the evolution of glial cells, biological insulators that surround axons and keep moving electrons on track. University College London biologist William Richardson's research group has speculated that glial cells evolved as modifications of motor neurons, not implausible given the patterns of gene expression that are common to the two. They further suggested that such glial cells could have immediately conveyed a large adaptive advantage by making it possible for prey to more rapidly escape their predators.[19] It could, of course, have been the other way around: Glial cells might have been an innovation that allowed vertebrates to *be* predators, giving them the fast reflexes they needed to make a meal out of a sluggish invertebrate.

Myelin insulation also made axons more energy efficient and made it possible for them to be packed closer together without cross talk, in turn making possible larger, denser brains. Pity the poor cephalopods, which never developed myelin. Octopi, for example, are among the few animals that can learn by imitation,[20] but without myelin, their evolution may have reached a ceiling.[21] If none of our ancestors had developed myelin it is doubtful that we would be able to read, hold a conversation, or drive a car. (One downside is that we are subject to disorders such as multiple sclerosis, in which myelin gets attacked by the immune system;[22] myelin may also be part of why recovering from strokes and other brain injuries is so hard.)

Second, with the development of myelin early in the vertebrate lineage came larger, denser brains with greater organization, including a three-part division into forebrain, midbrain, and hindbrain. Central to this transition was the fourfold duplication of a set of ancestral genes known as the *Hox genes,* named in honor of *homeotic mutations.* Homeotic mutations, first studied systematically in the 1890s, are unusual transformations in which repeated elements—such as vertebrae or the segments of a fly—sprout structures that ordinarily appear elsewhere—such as rare human cervical vertebrae that form ribs, or fly segments that grow extra wings in place of *halteres* (quasi-wings used for balance). Such mutations were eventually traced to a set of "selector genes"—*Pax6* (the master control gene for eyes that was described at the end of Chapter 4) is one of them—that govern the identities of particular segments. Such genes play a major role in patterning the basic body plan of everything from fruit flies to humans.[23] Fruit flies have a single set of *Hox* genes that is expressed selectively from front to back, each gene in a different segment. Through some lucky accident, some early vertebrate developed four copies of each.[24]

The quadruplication of the *Hox* genes made possible a whole new level of developmental precision. One set continued to give rise to the

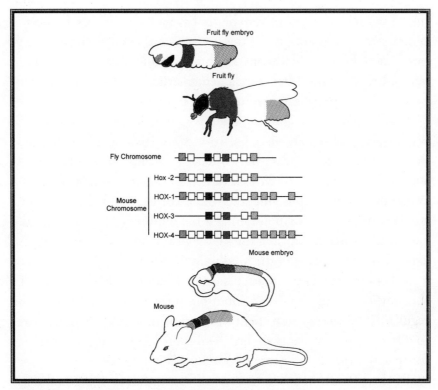

Figure 7.1   *Hox* genes in a fruit fly and a mouse
Illustration by Tim Fedak

basic segments of the body, but others were modified to guide the structure of the brain, the head, and the jaw, key elements in what makes vertebrates special. The head made possible the development of gills and specialized respiratory muscles, which in turn led the way to a more active lifestyle, and the jaws and the more complex brain made it possible for fish to make a living as a new kind of predator that bit with verve.[25]

By the time some of those early vertebrates crawled out of the water, perhaps 400 million years ago,[26] the rough structure of the mammalian brain was pretty much in place.[27] Mammalian brains differ from those of birds, fish, amphibians, and reptiles in many ways, but the differences are mainly in emphasis and detail. Humans have huge, highly articulated frontal lobes; birds have greatly enlarged basal ganglia. But all

five of the major classes of vertebrates have a basic division between a
central nervous system (placed toward the front) and a peripheral nervous system, between hindbrain, midbrain, and forebrain, and between
left and right hemispheres, fundamental structural divisions that support a massive specialization of neural labor.

That basic patterning of the brain is shaped in essentially the same
way in every vertebrate via regulatory genes (that is, genes that produce proteins that regulate the switching on and off of other genes)
such as *Otx* and *Emx*. As we saw earlier, *Emx2* shapes the balance between the hippocampus and the frontal cortex. Similarly, *Otx2* controls the balance of the midbrain and the hindbrain. If it is artificially
expressed (switched on) in a broader region than usual, the midbrain
expands at the expense of the hindbrain. If it is artificially restricted to
a smaller area, the opposite occurs: more hindbrain, less midbrain. In
this way, by modifying the distribution of the proteins (changing the
regulatory IFS that govern them), evolution tuned different creatures
to different environments, giving them more hippocampus if memory
for spatial location was more important (as it is for certain birds that
cache their food), or more forebrain if complex reasoning and decision making was particularly important (as it is for primates, including, of course, human beings).[28]

The evolution of this increased neural specialization of labor was,
like many other aspects of evolution, driven in part by possibilities
stemming from random gene duplications. For example, the amphioxus has a single copy of *Otx* and a single copy of *Emx*, whereas
most vertebrates have at least two copies of each—and thus greater
opportunity for fine-grained genetic guidance of what goes where.[29]
Although the copies are similar to each other, they have taken on different functions. For example, mice that have been engineered to
lack *Otx1* suffer from epilepsy and grow lots of abnormal neural connections, whereas mutants engineered to lack *Otx2* die early in embryonic development, apparently because they fail to develop the
precursor tissue that would lead to the forebrain, midbrain, and front
part of the hindbrain.[30]

Two vertebrate *Emx*'s where flies have one, two vertebrate *Otx*'s
where flies have one—such observations are pretty typical of a larger

vertebrate trend toward an elaboration of basic families of neural proteins, a trend that is especially clear in molecules such as semaphorins and ephrins that guide the growth of the brain's wiring. Vertebrates, for example, have five different families of semaphorins, whereas flies have only two.

Even these numbers pale, however, in comparison to what happened in the mammalian olfactory system a few hundred million years ago. Whereas our color vision depends mainly on having three separate types of color-sensitive cells in our retinas (each depending on a separate photopigment), mammalian olfaction depends on roughly 1,000 different receptors, each specialized to detect a particular ambient odor-causing chemical.[31]

Vertebrates also have some wholly new proteins—often formed by stringing together old elements of proteins in new ways—that look to be very important, such as *"reelin,"* named for *"reeler,"* a mutant mouse (mentioned earlier) that wobbles around as if it were drunk.[32] *Reelin* appears to contribute to both axon branching and the process of synapse generation. Its absence can lead to neuron degeneration,[33] and it may in some way be implicated in autistic people, who, according to one study, may have lower-than-ordinary levels of *reelin* expression.[34] Also prominent on the list of proteins that might provide clues into vertebrate brain functioning are nerve growth factors known as "neurotrophins"[35] (which, recall, are long-range signaling molecules that guide axons and influence neuronal cell survival), a set of receptors that are specialized for detecting nerve growth factors, and a special kind of cell adhesion molecule known as "protocadherins"[36] that may play a critical role in long-term memory or embryonic brain wiring.[37]

As the vertebrates split off into separate classes, each line specialized to a particular environment or niche. Amphibians adapted to a hybrid lifestyle, half aquatic, half terrestrial. Most birds adapted for flight, each species modifying the basic vertebrate plan in its own way.

Mammals developed a thin, six-layer cortical sheet known as the *neocortex*. That sheet, more than anything else, is what makes the minds of mammals so powerful. Less than 4 millimeters thick, it is divided into functionally specialized areas. In a mammal as simple as the hedgehog, there are about fifteen separate areas in the neocortex, including visual, somatosensory, and auditory areas, each divided into "primary areas" that receive input directly from relay stations such as the spinal cord and thalamus, as well as a motor area, a limbic region (likely involved in emotion), and a prefrontal area, likely devoted to planning and/or decision making. Each of these has counterparts in all mammals (save some cuddly Australians—marsupials and monotremes—that diverged somewhat earlier).[38] The exception that proves the rule may be the blind mole rat, which still has V1, otherwise known as the primary visual cortex, even though it cannot see.[39] (As you might expect, evolution doesn't just let V1 sit idle. Although the physical area is retained, its function is not; the blind mole's V1 has been adapted to process auditory information.)

Mammals with more complex brains, such as cats, dogs, monkeys, chimpanzees, and humans, follow much the same plan but have significantly bigger *cortices* (plural of "cortex") that devote more areas to each specialization. For example, both cats and monkeys have more than ten visual areas, thought to be specialized for different aspects of vision such as color and motion, and several somatosensory areas[40] (but not the same areas; after carnivores and primates diverged roughly 90 million years ago,[41] their cortical areas proliferated independently, perhaps in part as the result of independent gene duplications).

No matter how many cortical areas a mammal has, the basic layout is the same: The visual areas are closest to the back, the auditory areas are closer to the front (and closer to the ears compared to the visual cortex, which is closer to the midline between left and right), and the somatosensory areas are closer still to the front. In bigger brains, those areas move further apart, making room for new cortical areas in between.[42] Another trend is that as the brain gets bigger, it gets harder to stuff into the skull; all the wrinkles in a human brain are, in part, a product of nature's attempt to cram a big neocortex into relatively small space.

Mammals share not just an overall neural organization but also a system of developmental timing. Cornell psychologists Barbara Finlay and Richard Darlington have analyzed ninety-five neurodevelopmental

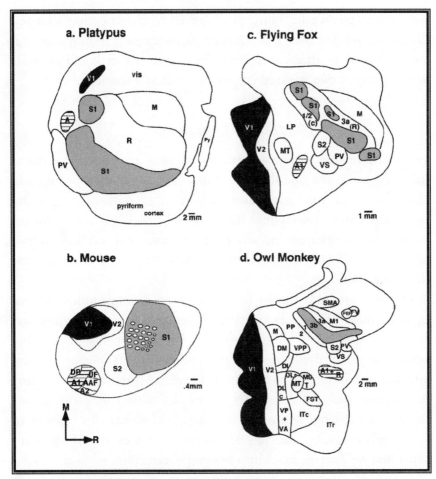

Figure 7.2   Cortical areas across the mammalian world
Illustration by Leah Krubitzer; reprinted by permission.

milestones—some within the cortex, some not—across nine species, from ferret to human and macaque monkey. Although the overall rate varied—brains develop a whole lot more quickly in hamsters and mice than in monkeys and people, the overall ordering of developmental milestones was virtually identical from one species to the next. For example, the fifth layer of the cortex always develops before the bundle of nerves that runs between hemispheres. The sequence is so regular that, given a single landmark in a particular species, Finlay and Darlington were able to predict the timing of all the rest with near perfect accuracy. Given the fact that layer V of the cortex is complete in a spiny mouse at twenty-two days after conception, for example, they could estimate within a day or two when its corpus callosum will develop.[43]

UNIQUELY HUMAN

For selfish reasons, I have a particular interest in one of those mammalian species, the one that communicates by rapidly opening and closing its vocal tract while doing funny things with its tongue and lips—a peculiar species that hides its nakedness in fabric and inexplicably spends time indoors reading books and watching television instead of gathering food or making babies—*Homo sapiens,* the loud-mouthed ape.

Humans are both similar to and different from our close animal cousins. Jared Diamond's point in the epigraph to this chapter was how similar we are to chimpanzees. In, for example, our body structures, our group dynamics, our perceptual systems, our aggression, and our sustained systems of maternal care, we surely have something in common with the chimp. Yet we are also plainly different. In our culture, in our language, and in our thoughts, we have the capacity to contemplate beauty, justice, calculus, and the meaning of life, concepts that we imagine no chimp has ever dreamed of.

Language, of course, must be a key to what makes our species unique. If learning is the genome's most powerful trick for moving beyond itself, language is arguably the most powerful tool for learning—the mother of all learning mechanisms and the single thing that most makes humans different. Language allows us to communicate information in ways that no other medium could match. It is clearly critical for the rapid transmission of culture, and it may even be a necessary component of some kinds of thought.

We often have the subjective impression that we think in words. Although scientific opinion is divided as to whether we really do sometimes think in words (or only think that we do), there can be little doubt that language is an important part of what makes us uniquely human. As a medium for communication, it allows elders to teach the young; as a medium for thought, it almost certainly helps us to store and retrieve information more efficiently, and it may even help us to reason more efficiently.[44] Charles Darwin himself suggested that some of our "certain higher mental powers" might be "the result of the continued use of a perfect language."[45]

Not everybody would agree that language is a medium for thought. Jerry Fodor, for example, has argued that language must be separate from a "language of thought," or "mentalese," because there is a slippage between language and thought.[46] There are, for example, thoughts that cannot be expressed with language and "tip-of-the-tongue" phenomena in which we know there's a word for something but can't quite come up with it.[47] The psychologist Lila Gleitman has argued, rightly in my view, that there is little good experimental evidence showing lasting cognitive differences between speakers of different languages. She suggested instead that "linguistic systems are *merely* the formal and expressive medium that speakers devise to describe their mental representations," and that "linguistic categories and structures [serve as] more-or-less straightforward mappings from a preexisting conceptual space, programmed into our biological nature."[48] Perhaps the most compelling argument for a difference between language and thought is one raised by Fodor and Steven Pinker: the contrast between sentences which are ambiguous (Does "bank" mean a financial institution or a riverbed in "Lloyd sat next to the bank"? Does "Everybody loves somebody" mean that each of us has our own particular soul mate, or that Jennifer Aniston is the object of everybody's affections?), and thoughts in mentalese, which presumably are *un*ambiguous. When I think to myself "Everybody loves somebody," I know what I mean.

But what do these arguments show? They show that there can be thoughts without language, but not that language plays *no* role in thought. Babies, monkeys, and aphasics (adults who have lost their ability to speak) all have thoughts, even if they cannot speak, and we all have thoughts that we can't put into words—emotions, sensations, and so forth—but that shows only that some thoughts are not linguistic, not that no thoughts are. I would like to suggest that language makes certain types of thought—including the kind of conscious, reflective reasoning you are engaged in right now—possible. Language has the potential to affect our thoughts in at least two ways: first, by "framing" the content of our ideas; second, by affecting what we remember. By "framing," I mean that language can, like a flashlight (or a hand gesture), point our attention in particular directions. When Henry Kissinger says that "mistakes were made," he aims to use language to divert us from the embarrassing question of who made the mistakes. Language is all about emphasis: As Lila Gleitman pointed out, saying "Meryl Streep met your sister" is entirely different from saying "Your sister met Meryl Streep."[49] As every good spin doctor must know, to frame a sentence is to frame a thought.

When it comes to memory, language's most obvious role is to help us "rehearse" information in our heads, as when you repeat a phone number to yourself. Remarkably, the number of digits a person can remember depends on what language they use—Chinese uses shorter words for numbers than English, which in turn uses shorter words than Welsh, and correspondingly, under carefully controlled conditions, Chinese speakers can remember more digits than can English speakers, who can in turn remember more than Welsh speakers. The effect is so strong that Welsh-English bilinguals who are generally more comfortable in Welsh still do better on digit memory when they are tested in English.[50]

Language may also facilitate thought by providing simple hooks for complex concepts. If you want to think about carburetors, for example, it may help to have a word for one. Specialists and aficionados have words like "grommets," "forceps," and "espresso" precisely because naming these objects helps to pick them out faster; it may be that once you have a word for a grommet, it's easier to learn what one does. Studies of stone-age cultures in Papua New Guinea by Jules Davidoff suggest that words may facilitate our memory for color by giving us category terms that allow us to carve up the spectrum;[51] it may be easier to remember that something is lavender if you have a word that distinguishes it from other colors.

Even more exciting to me is the possibility that language may also play a crucial role in long-term memory by providing a special way of *encoding* complex information. To understand this point, it helps to again think about how computers work. A computer's memory is made up of a long string of "bits" that can be either zeros or ones, but those zeros and ones mean nothing without an organizational scheme, or what computer programmers call a "data structure," a way of taking a particular set of zeros and ones to stand for a particular kind of entity, such as a number, a name, or an intensity for a picture element (pixel). What a particular computer program represents is a function of the particular types of data structures it can encode; some programs may store only pictures but not names, others the other way around. Language's greatest contribution may be in providing a data structure for storing relationships between entities and bits of information about those entities, or what linguists call subjects (say, "George W. Bush") and predicates ("felt that the American voters had misunderstood him"). Such a data structure might be a key to enabling humans to represent a uniquely broad range of thoughts.

But what about Pinker and Fodor's point about ambiguity? Do we really store thoughts as sentences? If so, why aren't our thoughts as ambiguous as our sentences? The answer, I believe, turns on a distinction between spoken sentences that truly are ambiguous and internal versions of those same sentences, which may not be. In internal versions of sentences, words such as "bank" might be marked for their meanings (bank$_1$ versus bank$_2$) and relations between words could be made explicit (for example, whether "somebody" means a unique soul mate or Jennifer Aniston in "Everybody loves somebody"). It is possible, then, that for a subset of our thoughts, an annotated kind of language—rather than a separate mentalese—could serve as a medium for long-term storage.[52] (If there is some kind of language-specific representational system, then it would stand to reason that children who have yet to learn a language would not have such a system and would therefore not be able to entertain the same range of thoughts as people who had acquired linguistic systems. We are not yet able to confirm or reject this conjecture.)

Whether language is a medium for thought or just for communication, its importance in our lives cannot be understated. Chimpanzees and orangutans have the rudiments of culture, but without language, and its capacity for rapidly transmitting—and perhaps encoding—a wide range of information, they will never have culture as rich as ours. But why is it that we have language, and our chimpanzee cousins, who share more than 98 percent of our genetic material, do not?

∞

When it comes to evolution, the question almost seems too easy. Dozens of eminent scientists have made proposals. Today, we have the aquatic ape hypothesis,[53] the language from gesture theory,[54] the theory that language arose from the neural machinery that evolved to control our muscles,[55] the theory that language came about as an accidental consequence of having bigger brains,[56] the theory that language is an extension of our capacity for representing space,[57] the theory that language evolved for the purpose of gossip,[58] and the theory that language evolved as a means of engaging in courtship and sexual display.[59] Probably more than a few of these theories have a grain of truth or two in them. Language does, for example, make gossip possible, and it couldn't have hurt our ancestors to know a little bit more about their neighbors than the next guy did. But, as linguists are fond of saying, languages do not leave fossils, and thus far, there has been very little evidence to tease apart one theory of the origin of language from the next.

Part of the problem is that we haven't yet figured out exactly what it is about the mind and brain that allows us to learn and use language in the first place. Until recently, most textbooks ascribed the ability to use language largely to two walnut-sized regions of the left hemisphere of brain known as "Broca's area" and "Wernicke's area."[60] According to the textbooks, Broca's area was important for grammar, Wernicke's for the meanings of words. Patients with lesions (induced by, say, strokes or gunshot wounds) in Broca's area would have trouble with syntax (for example, understanding the difference between

"The boy kissed the girl" and "The boy was kissed by the girl"), and patients with lesions in Wernicke's area would have trouble with naming familiar objects. Often implicit was the idea that each of these regions was dedicated solely to its own particular linguistic function—Broca's was the grammar area of the brain, Wernicke's the meaning area of the brain.

The only problem with this lovely story, now over a hundred years old, is that it's wrong. As scientists have discovered new imaging technologies (such as Positron Emission Tomography, or PET, and functional Magnetic Resonance Imaging, or MRI) to peer into the brains of unimpaired adults, they have found that Broca's area is indeed (in many experiments) active in syntactic processing[61] and that Wernicke's is active in understanding and producing words,[62] but they've also found that other areas of the brain participate in both kinds of processing and that neither Broca's nor Wernicke's area is restricted to purely linguistic functioning. Broca's area, for example, seems to be active not just in language but also in the comprehension of music (even by nonmusicians),[63] in imitation, and perhaps in motor control. In fact, the more people study Broca's area, the harder it is to discern exactly what it does. Meanwhile, it appears that syntactic processing engages other parts of the brain further to the front,[64] and perhaps "subcortical" areas that are not even in the evolutionarily recent neocortex;[65] studies of word-learning have implicated not just Wernicke's area but also visual areas, motor areas, and so forth. Rather than being confined to a single box in the head, our knowledge about words may be scattered across different regions of the brain.[66] Visual areas may play a role in our understanding of a word's visual properties, auditory areas in our understanding of the sound of a word, motor areas in our understanding of action verbs.[67] To top things off, even the classic studies of lesions turn out to be oversimplified; there are people with lesions in Broca's area that have normal grammar, and people with an intact Broca's area but damage elsewhere that have trouble with grammar. Ditto for the relationship between Wernicke's area and words.[68] Just as there is no simple one-to-one mapping between genes and brain areas, there is no simple one-to-one mapping between brain areas and complex cognitive functions.

Another approach to figuring out the neural basis of language might be to compare human brains with chimpanzee brains; if there were some obvious difference between them, researchers interested in the neural basis of language could start there. But the only immediately obvious difference between our brains and those of chimpanzees is in their size—the average chimpanzee weighs about 55 kilograms and has a brain of about 330 cubic centimeters, but the average human, who weighs only about 20 percent more, has a brain that is about four times larger.[69] Although that difference is important, it is unlikely to be enough by itself. Whales and elephants have brains bigger than ours,[70] but they do not have language. Big dogs have bigger brains than small dogs, but Great Danes are no smarter or more likely to talk than miniature schnauzers.[71] At 500 grams, the brain of a gorilla is over a hundred times bigger than that of pygmy marmoset (4.5 grams),[72] but I doubt a gorilla is even three times smarter. Intelligence (as measured by IQ tests) is only barely correlated with brain size.[73] Men have bigger brains (on average) than women, but (on average) women have better language skills.[74] Humans with unusually small brains can sometimes have language.[75] In short, size isn't everything. Having a normal, human-sized brain may be a prerequisite for language, but it is clearly neither necessary nor sufficient. Instead, as Barbara Finlay has suggested, it may be better to think of large brains as providing raw material that allows for the possibility of further evolution.[76] As my first animal behavior teacher, Ray Coppinger, put it, "When my very young son . . . [passed the brain size of an adult chimpanzee], he could tell you the batting averages of all the Red Sox players. Something else besides size is going on."[77]

But if it's not just size, what is it? At a gross level, our brains and those of the chimpanzee are structured in almost identical ways. We both have occipital cortices in the back of our heads wherein we analyze visual information; we both have brains split into left and right hemispheres, with interconnecting cables that run through the corpus callosum. We even (contrary to earlier conventional wisdom) share left-right asymmetries in Broca's and Wernicke's areas.[78] But a few differences, still quantitative rather than qualitative, have already been noted. For example, in humans the corpus callosum is proportionally smaller in comparison to the rest of the brain than it is in chimpanzees.[79] That means that humans have less communication between hemispheres, but at the same time (as measured by the amount of the white matter that contains neuronal connections) more communication within hemispheres[80]—a combination that may lead to (or be the consequence of) the degree of neural specialization that we might expect for language.

But once again, we have at most identified only a necessary condition, not a sufficient one. Elephants and whales continue the trend; their corpus callosa are even tinier in comparison to their colossal brains, and though having an intact corpus callosum is a good thing for learning language, it's not clear that it is absolutely necessary.[81] Another finding is that there may be some microscopic differences in part of Wernicke's area known as the *planum temporale*.[82] In both humans and chimps, the planum is larger on the left than on the right, but in the humans some substructures known as *minicolumns* are much bigger on the left than on the right.[83] Still, I doubt that these differences are enough; I wouldn't be surprised if there were many other important differences that our present-day microscopes just can't detect.

The upshot of all this is that at present we cannot simply point to a particular spot in the brain and say *this* is the language area, *this* is the neural circuit that makes this brain a uniquely human brain. To some scholars, these complex (and frankly unsatisfying) results challenge the innateness hypothesis because they spell the end of the "modularity" hypothesis, the idea that separate neural systems might be specialized for distinct neural functions.[84] To me, they suggest not that we should abandon modules (the Swiss Army Knife view of the brain) but that we should rethink them—in light of evolution. Nothing about the brain was built overnight; evolution proceeds, in general, not by starting over but by tinkering with what is already in place. As François Jacob famously put it, evolution is like a tinkerer who "often without knowing what he is going to produce, . . . uses what ever he finds around him, old cardboards, pieces of strings, fragments of wood or metal, to make some kind of workable object. . . . [The result is] a patchwork of odd sets pieced together when and where opportunity arose."

As we have seen, this is no less true for the brain than it is for the body. The ingredients—and genes—that make up our brains are, like the ingredients that make up the rest of our bodies, the product of evolution. New cognitive systems are patchworks and modifications of old. Specialized biological structures need not be, and perhaps never are, made up entirely, or even in large part, of wholly novel materials. Consider, for example, the differences between an arm and a leg. Thousands of genes contribute to each, but probably only a handful are special to the arm; if recent studies from mice and chicks are any guide, genes for everything from muscle proteins to bone marrow are likely to be pretty much the same in both.[85]

A language module may depend on a few dozen or a few hundred evolutionarily novel genes, but it is also likely to depend heavily on genes—or duplicates of preexisting genes—that are involved in the construction of other cognitive systems, such as the motor control system, which coordinates muscular action, or the cognitive systems that plan complex events. At the genetic level, figuring out what gives humans the unique gift of language will be a matter not just of finding out about those (perhaps relatively few) genes that are unique to people, but also a matter of finding out how those unique genes interact with all the others that are part of our common primate heritage. Essentially the same can be said at the brain level: Understanding language will be a matter not just of understanding unique bits of neural structure but also a matter of understanding how those unique structures interact with other structures that are shared across the primate order.

The ability to learn the rules of grammar, for example, may depend on circuitry for short-term memory that spans the vertebrate world, circuitry for recognizing sequences and "automatizing" (speeding up) repeated actions that is common to all primates,[86] and special circuitry for constructing "hierarchical tree structures" (see below) that is unique to humans. Our ability to acquire words may depend on a mix of long-term memory abilities that are found in animals and some special human facility, the details of which are not yet clearly understood.[87] (Consistent with this conjecture, amnesiac patient HM's inability to remember new facts extends to new words such as "granola" and "Xerox" that became common after his brain surgery.)

In short, from the perspective of evolution we should expect a language system to consist not of a single, brand new chunk of brain but of a new way of putting together and modifying a broad array of previously existing subsystems. Different parts of the brain probably are specialized for different functions, but most of those functions are likely to be shared *subcomponents* for computation, not complete systems for single-handedly solving complex cognitive tasks.

Data are scarce when it comes to humans, but animal models suggest that this way of thinking about things—that is, that neural machinery for new tasks evolved as novel combinations of mostly preexisting components—is on the right track. Take, once again, the example of fruit-fly courtship. The neural subsystems that support fly courtship have many of the properties one might expect from a mental module: They are fast (a fly doesn't need to reason on a blackboard to figure out its next move), automatic (a fly can do them without prior training or practice), and largely independent of other aspects of the fly's other cognitive processes (a male fly that catches sight of a receptive female is likely to drop everything and head straight for courting).

But the subsystems that support fly courtship are not, for the most part, unique to courtship. Many of the neurons, for example, that are involved as the courting fly rubs its wings together are neurons that are likely to also be involved in everyday wing movements that have nothing to do with courtship. The odor receptors that sniff for females also sniff for other smells. Only a relatively tiny of number of neurons—which may act as supervisors coordinating the actions of others—are likely to be *uniquely* involved in courtship; *fru* (the courtship-related gene I mentioned in Chapter 5) may work its magic mainly by guiding the connections between those supervisors and the rest of the many neurons that are required by—but not unique to—courtship. We may ultimately understand language in a similar way, as a powerful new combination of mainly old elements.

If the specialization-through-reconfiguration view that I am sketching is correct, we shouldn't expect there to be many mental disorders that uniquely affect a particular cognitive domain. If 95 percent of the genes involved in the circuitry for building language also participate in the construction of other mental capacities, the vast majority of genetically based disorders should be expected to have broad effects. Impairments to memory, for example, ought to impair language as well as other domains such as planning and decision making. Disruptions to the genes that code for metabolic enzymes might be expected to affect (in some cases) the entire brain, and disruptions to genes that code for receptor proteins might have effects wherever those receptors are found. It would be a rare disorder indeed that would affect only a single aspect of cognition.

To "anti-modularists," mental disorders that affect multiple domains are prima facie evidence that the mind is without modules. If some disorder affects both language and general intelligence, many assume that language is simply the product of general intelligence, not an independent entity. But a finding that a given disorder affects two behaviors or neural structures doesn't mean that the two are identical—it just means that they are built in a similar way.

Even professionals sometimes seem to be misled by an apparent analogy with brain damage and the "logic" of "double dissociations." If a single brain region is heavily involved in two processes, it is reasonable to assume that the two domains involve some of the same computations. In contrast, a single gene that is involved in two different processes does not necessarily show that the same cognitive functions are produced by the same bit of neural structure. Because of the richness of gene regulation, a single gene may be used multiple times in the service of radically different functions. The protein product of the gene *microphthalmia-associated transcription factor (MITF)*, for example, participates in eye formation, blood cell formation, and pigmentation.[88] If the neural substrates of language are built using some of the same genetic cascades as the neural substrates of general intelligence, we shouldn't be surprised that some disorders affect both.

Conversely, if complex skills draw on many different underlying components, we shouldn't be surprised to see them vulnerable in several different ways. The ability to read, for example, depends on linguistic knowledge (that is, what words do and do not sound like: Is "mave" more likely to rhyme with "have" or "gave"?), visual prowess (the ability to detect the critical left-to-right difference between a "d" and a "b," strange in the context of a world where left-to-right differences are otherwise often unimportant), and the ability to connect those funny visual squiggles to linguistic sounds (in ways that are, once again, purely arbitrary; "P" is pronounced like a "p" in English but an "r" in Greek, and never mind all the different sounds that the string of letters "ough" can signify in English). The visual system needs to take special steps to make two eyes focus on a single page just a few inches from the face (weird, again, in the context of the natural world, where much of what we need to see is more distant), and since there is no way to master the sounds of a language if they can't be heard properly, the linguistic system depends on the auditory system working properly.

So it is no wonder that a disorder such as dyslexia (a severe, highly heritable reading disorder not associated with other cognitive, memory, or motivational impairments) may arise in many different ways in different people, sometimes as a result of poor audition, sometimes as a result of poor vision, sometimes as a result of impairment in the linguistic system.[89] According to one study, about two-thirds of dyslexics see words as blurry, or jumping around on the page, and many are actually helped by closing one eye (probably because their problem has something to do with the problem of coordinating the two eyes as the eyes rapidly scan across a printed page).[90] Other dyslexics seem to have perfectly normal vision but may have problems with hearing the quick, subtle, spectrographic changes that characterize speech,[91] and still others may have entirely normal sensory ability but significant problems with phonology, the part of language that concerns the sounds of words. Nature has no commitment whatsoever to a principle of one disorder, one cause. Just as disorders of the body can be caused in dozens of different ways (for example, malnutrition from missing enzymes, disordered organs, even bad teeth), disorders of the mind may result from many different underlying aberrations.

The real question—as yet unanswered—might be whether there are *any* pure, genetically derived disorders that affect only a single cognitive domain such as language. The only speech and language disorder that has been decisively tied to a particular gene—the *FOXP2* gene, which I'll describe more extensively later in this chapter—is clearly not special to language. For now, there is only tantalizing evidence—not yet tied to specific genes—that a few rare disorders might indeed uniquely affect certain aspects of language. Some versions of dyslexia that stem from focused trouble with phonology are a good example. Another is "G-SLI" (Specific Language Impairment that is grammar-specific), a language disorder studied by the psychologist Heather van der Lely that seems to specifically affect aspects of grammar, such as understanding passive sentences ("The boy was kissed by the girl") and producing past-tense versions of novel forms ("This is a man who knows how to wug. Look there, he just did. He just ____").[92] If such disorders can be tied to particular genes,[93] they could give us valuable clues—not because the genes that would be impaired in such disorders would be the only genes involved in language, but because they would suggest insights into those parts of the neural substrates of language that are unique.[94] Even if most disorders affect many domains, and many domains are affected by many disorders, it's far too early to count the modular theory out.

If language is an amalgam of off-the-shelf components shared with other primates and a small number of human-specific components, which human-specific components might have made the difference? One possibility is that something special about human *social* cognition might have been significant. For example, both Michael Tomasello, an expert on primate social cognition, and Paul Bloom, a psychologist whose principal interest is how human children learn words, have contrasted chimpanzees' apparent ineptness in understanding the goals and intentions of others with human children's relative facility for such things. For example, if an experimenter hides a piece of food in one of two opaque containers and points to the one on the right (or looks at it, or taps it), you might expect the chimp to go for the one that the experimenter is indicating, but even after dozens of trials, some chimps just don't seem to get the game,[95] a sign that they may not understand enough about the intentions of others to get into the communication game. Bloom wrote, "The failure to appreciate the representational intentions of other people . . . is so extreme [in chimpanzees] that it entirely precludes word learning."[96]

Learning language does not, of course, literally require you to be able to look within the soul of the boy sitting next to you (to paraphrase Woody Allen's old joke about what he did at a metaphysics final that got him thrown out of NYU). But it surely does help if you can make good guesses about what other people are likely to talk about, and that comes from a good sense of human psychology and the faith that the psychology of other people is probably not altogether different from your own. (You can never know for sure. You could be the only truly conscious person; the rest of us all could, for all you know, be zombies or robots. As the logician Raymond Smullyan once put it, "Sure I'm a solipsist. But then again, that's just one man's opinion.")

From the time children start learning the meanings of words, they recognize that it is important to take the beliefs of others into account. One of the first experiments to make this clear was conducted by Dare Baldwin as part of her Ph.D. dissertation in 1990.[97] Baldwin sat sixteen-month-olds in front of an interesting toy and said, "Look at the *toma*." The test was whether a child would think that the word "toma" referred to the toy that she herself was looking at (for example, an extendable green periscope)—or a different toy that the experimenter was looking at (a disc surrounded by yellow suction cups). Shattering Piaget's claim that toddlers are egocentric to the point of being unaware of the perspective of others, the sixteen-month-olds consistently took "toma" to refer to the toy that the experimenter was looking at.

By the time children are four years old[98] (maybe even earlier, according to some new studies by Renée Baillargeon[99]) they know that other people can have *false* beliefs. Try this game with your four-year-old neighbor. With her father looking on, show her some cookies, put the cookies in a cookie jar, and then ask her father to leave. While he is out of the room, move the cookies into the refrigerator; now ask your young neighbor, "Where will Dad look for the cookies?" If your neighbor has mastered the notion of "false beliefs," she'll guess (perhaps with a giggle) that Dad will still look in the cookie jar. If she has not yet mastered the concept, she may think that since *she* now knows that the cookies are in the refrigerator, Dad will know, too. Chimpanzees seem to be lousy at nonverbal equivalents to these tasks,[100] and as I mentioned above, even pointing seems to throw them for a loop. As far as we can tell, chimpanzees are (except perhaps when it comes to competing for food[101]) lousy intuitive psychologists, a fact that could indeed play into their troubles in learning language.

But that's a far cry from saying that mind reading or social savvy is *necessary* for the acquisition of language—it's much easier for me to guess what you're talking about if I can guess what you might be thinking about, but I'm skeptical of any theory that holds that social intelligence is essential to acquiring language. Some high-functioning autistic people, with clearly impaired social cognition, are able to acquire fluent or near-fluent language, while chimpanzees like Kanzi clearly enjoy social interaction yet lack the ability to acquire full-blown language. Dogs arguably have even better social skills and are apparently able to understand eye gaze better than chimpanzees, yet they, too, rarely master the ability to understand more than a few words or simple commands.

Another critical factor may be the almost magical ability of humans to combine simple elements into more complex ones that *can in turn serve as elements in further combinations,* an idea sometimes referred to as "recursion." If you can think about a ball, you can think about a big ball, and if you can think about big ball, you can think about a big ball with stripes, a big ball with stripes that lies on the beach, and so forth. In a computer program, this would (in part) be a matter of "representational format," of how memory is organized. Everything that is stored in a computer must be assigned to a particular category—a name, a phone number, a picture, or what a computer scientist might think of as a tree structure.

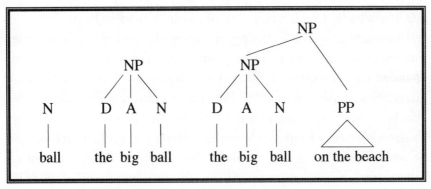

Figure 7.3 Three syntactic trees, illustrating recursive composition

If I had to bet on one thing that might have given rise to language, that talent for recursion would be it.[102] (Marc Hauser, Noam Chomsky, and the cognitive scientist Tecumseh Fitch recently argued for a similar position.[103]) Recursion is a key to what allows languages to describe so many things in so many different ways. Picture how disorganized your computer files would be if you couldn't make use of folders (or what die-hard Unix users refer to as "directories"). From the perspective of the programmer, folders (or "directory structures") are just a special way of organizing information, but one that radically improves the user's access to that information. The evolutionary addition of a new data structure for recursion—which is mathematically close to what programmers use to store folder structures—could be tiny from the genetic perspective, but profound in its consequences for communication and thought.

The ability to freely learn new words might be similarly powerful. Others have pointed to a set of special mental abilities for combining, analogizing, and integrating conceptually disparate elements (which themselves could give rise to—or originate from—recursion in language),[104] an ability to apply machinery from the motor control system to the problem of language,[105] or special changes in the human vocal tract (an idea now thought to be less plausible).[106]

Of course, there's no reason to think that language arose from a single innovation. A number of linguists and psychologists have proposed accounts of the evolution of language in which it may have arisen gradually over a series of steps, with abilities to learn words, to learn grammar, and so forth coming on one by one. If this is the case, our ancestors may have spoken "protolanguages" that might have sounded like newspaper headlines or telegrams. ("Human learn language. Neanderthal worry.")[107] For example, a thoughtful account by Ray Jackendoff sketched a series of twelve steps that could have led to the development of modern language, ranging from the development of an unlimited class of symbols to the development of a system of inflections (such as -ing and -ed) to convey semantic relations (like pastness).[108]

Language is, without a doubt, as complex as Jackendoff suggested; each of the steps that he described corresponds to something real about language. But I am not fully convinced that each corresponds to a separate, language-dedicated step in evolution, for two reasons. The first is that although there is a long tradition in studies of evolution of trying to account for change through gradual steps, I believe that the evolution-through-gradual-change assumption must be re-evaluated in light of the kinds of genetic mechanisms described in this book. Genetic changes presumably are gradual—mutation by mutation, duplication by duplication. But since, as we have seen, a single change can induce a new cascade or block an old one, phenotypic changes need not be gradual. A single mutation (or duplication) can have a large effect on the phenotype, the shape and behavior of an organism, as when a photoreceptor duplication leads to a new ability to see color. If language arose de novo, it would, I suspect, have to go through a long series of gradual steps, but if language arose by a novel combination of existing elements—such as neural structures for memory, the automatization of repeated actions, and social cognition—it is possible that it could have developed relatively quickly. The second is that, although I surely agree with Jackendoff that there is a considerable genetic contribution to our ability to acquire language, I am not as convinced as he is that so much of the machinery for language is special to language. He could be right, but he could be wrong; perhaps only a relatively small part of the language faculty depends on human-specific machinery. It is to my great chagrin as a lifelong student of language acquisition that the challenges in understanding the ability to learn language are so difficult that we still just don't know.

∞

Although I do not expect that scientists will in the next few years come to completely understand the neural or cognitive basis of language, let alone its evolution, I do think that genes themselves may soon give us some important insights.

At first blush, the genome may seem like an unlikely source of evidence. We diverged from chimpanzees only recently, in the evolutionary scheme of things—perhaps only about 4 to 7 million years ago,[109] just an instant compared to the 85 or so million[110] years in which there have been primates and the 3.5 to 4 billion years in which there has been life on the planet.[111] And, as Jared Diamond, among others, has emphasized, our genomes are not that different from those of chimpanzees.

This surprising fact was first discovered in 1975 by Mary-Claire King, then a graduate student at Berkeley, and Allan Wilson, already well known for his suggestion that changes in DNA over time could be used as a "molecular clock."[112] The molecular clock, which is based on the assumption that DNA mutates at a relatively constant rate, uses changes in DNA sequence to estimate when various species diverged. For example, if two species are closely related, their hemoglobin sequences will be quite similar; if they are distantly related, their hemoglobin sequences will be less similar. At the time of King and Wilson's work, it was not practical to spell out the sequence of nucleotides in large quantities of DNA, but it was possible to get an estimate of how similar two strands of DNA were by seeing how strongly the two strands stuck together. If two strands of DNA are identical, they form a strong bond and it is hard to get them to break apart; if they are different, they form a weaker bond and it is easier to get them to break apart. As a rough rule, every percentage point of difference leads to a degree (Celsius) change in their "melting point."[113] (The so-called melting point is really the point at which DNA strands separate, well below the point at which DNA would change from a solid to a liquid.)

King, who later became famous for her role in the discovery of the breast cancer gene BRCA1,[114] generalized Wilson's techniques and applied them to the problem of understanding what makes humans so special. Nobody expected that hemoglobin would be that different from humans to chimpanzees, but King and Wilson found that a wide range of other genes and proteins were also quite similar between the two species. Regardless of how they tried to measure the difference, whether they compared nucleotides or amino acids, the differences between humans and chimpanzees were tiny, no more than 1 or 2 percent, and their startling results were quickly picked up in the media. (Carl Sagan had a hand in popularizing them, writing with Ann Druyan in *Shadows of Forgotten Ancestors* that "By these standards, humans and chimps are about as closely related as horses and donkeys, and are closer relatives than mice and rats, or turkeys and chickens, or camels and llamas. . . . If we want to understand ourselves by closely examining other beings, chimps are a good place to start."[115])

And soon the King and Wilson results were replicated by other scientists. Today, the methods are more sophisticated—contemporary studies rely not on the melting trick of hybridizing DNA strands from different species but on computerized nucleotide comparisons of sequenced DNA—but the results remain the same.[116] Take almost any stretch of a hundred chimpanzee nucleotides and chances are that there will be a comparable stretch in your genome—and only one or two nucleotides will be different.

As King and Wilson suspected (but couldn't then know), the differences are mostly elsewhere—in the regulatory IF regions. As we have seen, even tiny differences in regulatory regions can lead to large differences in behavior, impairing memory or making one species more sociable than the next. A disproportionately large number of the differences between our genomes and those of chimpanzees are found in what are called *CpG islands,*[117] stretches of DNA that are strongly associated with the regulatory IF sequences that govern when genes are expressed.[118] In fact, although less than 1 percent of other sequences differ, roughly 15 percent of all CpG islands differ from chimp to man. What this tells us is that we are built out of the same proteins as our chimpanzee cousins, but there are important differences in how those proteins are organized. The protein template THENS of most genes are nearly identical—but those proteins are regulated in significantly different ways, and it may well be that it is this new regulation of old proteins that led to our ability to speak and to acquire culture.

The ability to speak, and the corresponding ability to understand language, because it likely builds on many other systems already in place in the primate line (such as memory and skill-learning), probably didn't arise overnight from a single mutation. But it may not have taken an enormous number of steps, either, especially if they involved changes in regulatory regions (particularly those arising from duplication at the top of complex cascades). As we start identifying genes that are implicated in language disorders, we can begin to ask how the genes that are involved in language relate to our other genes. Are they duplications (with minor divergences) of genes involved in motor control systems? Are they adaptations of genes involved in systems for mentally representing space?

Although the media often talk about "genes for language," most of the genes that are involved in language won't be unique to language. Language (and whatever else is special about the human mind) comes not just from the 1.5 percent of genetic material that separates us from the chimpanzees, but also from the 98.5 percent that is shared (and the ways that that 98.5 percent are influenced by the 1.5 percent that differ).

I suspect that researchers will find that most of what makes up the gift for gab is a product of roughly the same kinds of neural tissue that support other mental functions. The nerve cells that support language, like the nerve cells that support vision and motor control, are probably mostly made up of the usual stew of axons, dendrites, and myelin, albeit arranged in importantly different ways. Language, like any other cognitive process, makes use of memory, and scientists are likely to discover that it does so using roughly the same kinds of molecules as memory for, say, faces or events. If language came onto the scene relatively quickly by evolutionary standards—a few million years at most, perhaps a lot less, in a space of less than 100,000 years,[119] either way a brief instant in comparison to the hundreds of millions of years it took to evolve eyes)—it is because much of the genetic toolkit for building complex cognition was already in place.

In effect, the problem of finding genes for language becomes a problem of finding needles in a haystack. As we have seen throughout this chapter, recent evolutionary innovations tend to build on older evolutionary innovations by reorganizing and modifying existing structures, not by starting from scratch. I suspect that thousands of genes will be involved in the development (and maintenance) of the parts of the brain that support language, but that no more than a couple hundred of them will turn out to be unique to language. To understand the origin of language will be to understand how a relatively small set of new genes coordinates the actions of a much larger set of preexisting genes.

The first steps toward that project may already be under way. In 2001, a team of British geneticists, led by Simon Fisher, Anthony Monaco, and their student Cecilia Lai, uncovered the first gene to be tied decisively to an impairment of speech and language.[120] That gene, *FOXP2,* was discovered during the course of an investigation into a rare, single-gene speech and language disorder that is prevalent in a particular British family known as the KE family. The concordance between the gene and the disorder is perfect: Every member of the KE family that has a disrupted version of *FOXP2* has the disorder, whereas none of the members of the family that have normal versions of the gene are so impaired. Those with the disorder have trouble with, among other things, past tense verbs, repeating what psychologists call nonwords (for example, *pataca,* a word that sounds like English but isn't), and understanding spoken language.[121]

As the developmental neuroscientist Faraneh Vargha-Khadem has shown, the deficit is not restricted purely to language; afflicted members of the family also have trouble controlling sequences of movements with muscles of the mouth and face, and they have been known to have difficulties with instructions such as "Stick out your tongue, lick your upper lip, and smack your lips."[122] The value in *FOXP2* is not so much that it is "the language gene"—as I have been at pains to argue, I expect there is no such thing—but that it might serve as an entrée to unraveling the many other genes that are also involved in the cascades of genes that build language. *FOXP2* is not unique to people, and it is not unique to the brain, or even to language. Mice[123] and chimpanzees have it, and it is expressed in the lungs[124] as well as in the brain. But the human version of it is different in ways that may turn out to be important for understanding the genetic basis of language.

Although the human and chimpanzee versions of the 715-amino-acid-long protein are nearly identical, the human version contains two amino-acid changes that may play an important role in how the gene regulates its targets (a threonine that has been changed in the human lineage to an asparagine, and an asparagine that has been changed to a serine).[125] Two changes may not seem like much, but the same changes have been found in all humans studied, including Africans, Europeans, South Americans, Asians, and Australians, and the asparagine-changed-to-serine is found in no other primate, while the threonine-changed-to-asparagine has not been found in any of the twenty-nine nonhuman species that have been studied, from seals and chickens to monkeys and chimpanzees. Mathematical modeling suggests that the gene changed sometime in the past 100,000 to 200,000 years,[126] consistent with many estimates of when language itself may have evolved.[127]

For now, all we know for certain is that *FOXP2's* protein product is a regulatory protein, one of those genes that can trigger the action of others; it may or may not play a key role in language. But by tracing the cascades of other genes that it interacts with, how those genes vary across the animal world, and what role they may play in other cognitive systems, there's hope that someday soon we may be able to begin to piece together the molecular history of our gift for language.

8

PARADOX LOST

More with Less
—Proverb

FOR ALL THAT I have said about babies, genes, and brains, I still have not resolved the Two Paradoxes introduced at the end of the very first chapter: How is it that innateness and flexibility coexist? And how is it that a genome with far fewer than 100,000 genes can guide the growth of billions of neurons (not to mention the trillions of connections between those neurons)? With the tools of embryology firmly in our grasp, we can at last begin to answer these questions.

BRAIN PUTTY, REDUX

At the core of our story has been a tension between the evidence that the brain can—like the body—assemble itself without much help from the outside world, and the evidence that little about the brain's initial structure is rigidly cast in stone. The ocular dominance columns can form in the dark, yet they can be radically altered if one eye is blind from birth. The "visual parts of the brain" respond to visual input—except when they don't. Language is on the left—except when it isn't.

To an earlier generation of scholars, the evidence for innateness and the evidence for flexibility seemed almost irreconcilable. Most scholars simply focused their attention on the stream of evidence that they were more impressed with. Nativists gathered examples of what a child could do without the benefit of experience; empiricists gathered examples of how much the basic structure of the brain could change in response to challenges from the environment.

Both sides have their points. The brain is capable of awesome feats of self-organization—and equally impressive feats of experience-driven reorganization. But the seeming tension between the two is more apparent than real: Self-organization and reorganization are two sides of the same coin, each the product of the staggering power of coordinated suites of autonomous yet highly communicative genes. Just as a group of well-trained musicians can play a traditional piece or improvise a new one, suites of genes can play their standard tune or develop a new variation on a theme, as circumstances require.

Now that we have learned how genes truly work, and how it is that they contribute to the process of brain development, we can see that both development and reorganization—redevelopment—flow naturally from the bipartite nature of genes. By providing both a template for a protein and a guide to when and where that protein should be built, a gene contributes both a tune and a sense of the circumstances in which that tune ought to be played.

We have seen that the impressive plasticity of the brain is in many ways just a special case of the impressive plasticity of the body. The ability of young, uncommitted cells to adapt to new surroundings is just as true of wannabe eye cells transplanted into the stomach as of wannabe somatosensory neurons transplanted into the visual cortex. What makes both possible is the process of gene regulation, a process which, by its very nature, makes structure-building context-dependent.

The biological mechanisms that give rise to developmental flexibility are many and varied. Construction (and reconstruction) controlled by systems of gene regulation provide one kind of flexibility. Another key mechanism is redundancy: Just as a new Boeing 777 has three complete sets of computers to make sure that all goes well even if one set of computers fail, nature has backups—and backups for its backups. More than that, the body is filled with mechanisms for self-repair.

Some seem to arise directly from the mechanisms by which the fates of genes are specified. What controls the fate of a cell is, as we saw in Chapter 4, largely the patterns of gene expression within the cell, a matter of which genes are on and which are off. A presumptive eye cell can become a stomach cell because it contains the genes for building proteins appropriate to a career as a stomach cell (virtually every cell contains a complete copy of the genome). Moreover, the regulatory IF regions of those genes are responsive to cues to location[1] or molecules (such as pepsin) that are commonly found in the stomach, and the same, presumably, holds for a somatosensory cell transplanted to the visual cortex. By making the function of a given cell the product not just of that cell's individual history but also of the signals it receives, the genome guarantees that each cell will have a measure of flexibility—a kind of flexibility that comes almost for free.[2]

Other kinds of developmental flexibility require a little bit more from the genome, a cue that something needs to be repaired. When a salamander regrows a lost limb, much of the regrowth depends on the genes for building limbs in the first place, but there is also a system for inducing that regrowth, one that depends on a signal—probably retinoic acid—that spurs the limb growth cascades on to an encore performance.[3] And, as one would expect, the same signals may be at work in the brain: At least in the confines of a Petri dish, retinoic acid can induce neurogenesis.[4]

Although we are a long way from understanding all the details of how the nervous system can regenerate itself (retinoic acid is just one of many contributing factors), it is already clear that genes participate in every step of the process. Consider, for example, the lowly flatworm, unlucky in the IQ sweepstakes but one of the world champs when it comes to regeneration. Sever a flatworm at its head, and its head region will grow a new trunk and tail (and the tail and trunk will grow a new head, complete with a new brain). Francesc Cebrià, Kiyokazu Agata, and their collaborators at the RIKEN Centre for Developmental Biology in Kobe, Japan, have found that about a dozen different genes are upregulated (expressed more) during regeneration, each at a particular time, some in the early stages of regeneration, others in middle and later periods. As in so many other aspects of development, gene expression is under tight temporal control.

The worm is so hell-bent on regenerating lost brain tissue that it will do so anywhere that a gene known as *ndk* is not expressed. When the RIKEN lab interfered with *ndk* (named from the Japanese *nou-darake,* meaning "brain everywhere"), regenerating flatworms grew brain tissue all over their bodies.[5] Manipulate the genes, and you manipulate the ability to reorganize in response to injury. (This result is no sterile academic exercise. *Ndk* is closely related to a human gene *FGFRL1,* and there is a good chance that what happens in the worm will have implications for the regeneration of human neural tissue.)

Comparisons between closely related species underscore the point that developmental flexibility depends on having the right genes. For example, goldfish and frogs are blessed with a remarkable ability to recover visual function even after severe damage to their optic nerves (bundles of axons that run from the eye to the part of the midbrain known as the *tectum*). When, as part of research to help humans recover from injuries to their brains and spinal cords, experimenters crushed the optic nerves of fish and frogs, complex connections from the retina to the tectum regenerated in just a few months. In contrast, the corresponding connections do not regenerate properly in ornate dragon lizards *(Cteniphorus orgnatus).*[6] The system for developing connections from the eye to the midbrain is largely the same in the dragon lizard as in the frog, but the lizard's regenerating axons make many mistakes on their way to the visual part of the brain, entering nonvisual areas and even finding their way to the wrong eye. But the biggest mistake is with the axons that *do* make it to the normal visual target: They fail to sort themselves properly once they arrive, creating a scrambled map rather than a normal, well-ordered map of the visual world. The kaleidoscopic view the lizards wind up with renders them effectively blind in the experimented-upon eye. As neuroscientist Jenny Rodger has shown, the problem is a matter of gene expression: Axon guidance molecules and other growth-related proteins were expressed improperly.[7]

According to some developmental biologists, the ability to regenerate from damage is almost a default, the natural state of developing biological systems—something that is occasionally lost rather than something to be specially added in. (The myelin that I was praising in Chapter 7 may actually be the culprit. At least three different myelin proteins inhibit neural regeneration[8]—a heavy price for the ability to think and reason.[9]) Organization is not in conflict with reorganization; instead, both are natural consequences of a system of growth guided by genes.

It's no accident that developmental flexibility evolved. In mammals, which invest a large amount of energy and resources into each offspring, mechanisms that increase the chance that a given embryo will be viable are highly advantageous. Morning sickness, for example, may be nature's way of screening a mother's diet, eliminating potential toxins.[10] Development through gene regulation is a way of buffering a growing embryo against the accidents and vicissitudes of embryonic life. As Oberlin College biologist Yolanda Cruz has put it:

> In a rapidly growing embryo consisting of cells caught in a dynamic flurry of proliferation, migration, and differentiation, it would be desirable for any given cell to retain some measure of developmental flexibility for as long as possible. Such would enable an embryo momentarily disabled by cell cycle delay, for instance, or temporarily compromised by loss of a few cells, to compensate for minor disruptions and resume rather quickly the normal pace of development. It is easy to see how such built-in [flexibility] could contribute to the wide variety of procedural detail manifest in nearly every phase of mammalian embryogenesis.[11]

Regulative development also increases the odds that the parts of a given organism—combined as they are from maternal and paternal contributions—will work together. To take a fanciful (and oversimplified) example, an organism that got its biceps length from its tall father but the length of corresponding nerves from its short mother might wind up with a biceps nerve too short to reach the brain (to say nothing of arms too short to box with God). In the individual genetic shuffling that gives rise to a given embryo, and in the repeated shuffles that collectively produce evolution, organisms that can specify structure in relative terms will be far more likely to thrive; organisms that specified structures in absolute terms might be collections of mismatched parts

that would never make it out of the womb alive. Regulative development, and the flexibility to which it gives rise, is not the antithesis of
innateness but its guarantor. Flexibility is not the enemy of natural
complexity but its quality-assurance chaperone, a guardian that ensures
that the product gets built right—and the inevitable consequence of a
biology that uses positional cues to assemble itself dynamically.

COMPACT GENOME

The problem of using a small amount of information to describe a complex organism is a bit like a problem that computer scientists face when
they want to store and transmit information as efficiently as they can.
Back in the days when computer memories were measured by the kilo-
byte instead of the gigabyte, programmers developed ways of *compress-
ing* the information in a picture or word-processing document so that
they would use less space on a hard disk. Even today, when computer
memory is relatively cheap, building a better compression technique is a
good way to get rich, because the bandwidth of the Internet—the
amount of data that can be moved at a given moment—is not infinite.
Nobody likes to wait for a web page to load, and the more a webmaster
can compress a picture or a movie, the faster it will transmit over the
web. Better compression means more customers for your dot.com.

All compression schemes rely in one way or another on ferreting out
redundancy. For instance, programs that use the GIF format (pronounced "jiff," to remind us of the peanut butter[12]) look for patterns
of repeated pixels (the colored dots of which digital images are made).
If a whole series of pixels are of exactly the same color, the software that
creates GIF files will assign a code that represents the color of those
pixels, followed by a number to indicate how many pixels in a row are
of the same color. Instead of having to list every blue pixel individually,
the GIF format saves space by storing only the code for blue and the
number of repeated blue pixels, two numbers that can stand for fifty or
a hundred pixels. When you "open" a GIF file, the computer converts
those codes back into the appropriate strings of identical bits; in the
meantime, the computer has saved a considerable amount of memory.

So-called "vector-based formats"—first brought to the attention of my generation by the 1980s video game *Asteroids,* and now familiar from programs such as MacDraw and Adobe Illustrator—describe images in terms of their geometry, lines, curves, rectangles, and so forth. In essence, rather than storing a "bitmap"—a long list of ones and zeros that corresponds to colors of pixels—vector-based formats store a recipe for reconstructing the original image by redrawing the lines and curves that go into the image. Computer scientists have devised literally dozens of different compression schemes, ranging from JPEGs for photographs to MP3s for music, ZIP files for programs, and highly specialized (and less familiar) systems for storing fingerprints[13] and faces,[14] each designed to exploit a different kind of redundancy. In all cases, the general procedure is the same: Some end product is converted into a compact description, and a "decompressor" (my favorite is Stuffit Expander) reconstructs it.

Biology doesn't know in advance what the end product will be; there's no Stuffit Compressor to convert a human being into a genome. But the genome itself is very much akin to a compression scheme, a terrifically efficient description of how to build something of great complexity—perhaps more efficient than anything yet developed in the labs of computer scientists (never mind the complexities of the brain, there are trillions of cells in the rest of the body, and they are all supervised by the same 30,000-gene genome). And although there is no counterpart in nature to a program that compresses a picture into a compact description, there is a natural counterpart to the program that decompresses the compressed encoding, and that's the cell. Genome in, organism out. Through the logic of gene expression, cells are self-regulating factories that translate genomes into biological structure.

To the extent that genomes can be thought of as compressed encodings of biological structures, they are spectacularly efficient. All the trillions of cells in the human body—not just the tens of billions in the brain—are guided in one way or another by the information contained in 30,000 or so genes.[15] The best high-quality set of pictures of the body—the National Institutes of Health Visible Human Project, a series of high-resolution digital photos of slices taken from volunteer Joseph

Paul Jernigan (deceased)—takes up about 60 gigabytes, enough (if left uncompressed) to fill about 100 CD-ROMs—and *still* not enough detail to capture individual cells. The genome, in contrast, contains only about 3 billion nucleotides, the equivalent (at two bits per nucleotide) of less than two-thirds of a gigabyte, or a single CD-ROM.

How does the body push the comparatively tiny genome so far? Many researchers want to put the weight on learning and experience, apparently believing that the contribution of the genes is relatively unimportant. But though the ability to learn is clearly one of the genome's most important products, such views overemphasize learning and significantly underestimate the extent to which the genome can in fact guide the construction of enormous complexity. If the tools of biological self-assembly are powerful enough to build the intricacies of the circulatory system or the eye without requiring lessons from the outside world, they are also powerful enough to build the initial complexity of the nervous system without relying on external lessons.

The discrepancy melts away as we appreciate the true power of the genome. We could start by considering the fact that the currently accepted figure of 30,000 could well prove to be too low. Thirty thousand (or thereabouts) is, at press time,[16] the best estimate for how many *protein-coding* genes are in the human genome.[17] But not all genes code for proteins; some, not counted in the 30,000 estimate, code for small pieces of RNA that are not converted into proteins (called microRNA), or "pseudogenes," stretches of DNA, apparently relics of evolution, that do not properly encode proteins. Neither entity is fully understood, but recent reports (from 2002 and 2003) suggest that both may play some role in the all-important process of regulating the IFS that control whether or not genes are expressed. Since the "gene-finding" programs that search the human genome sequence for genes are not attuned to such things—we don't yet know how to identify them reliably—it is quite possible that the genome contains more buried treasure.

Moreover, the sheer number of genes underestimates the amount of information in the genome because a single gene can have multiple functions associated with multiple regulatory regions. Signaling genes, for example, use multiple regulatory regions to set up different gradients in different parts of the body. *Fru,* the fruit fly courtship gene, actually comes in different forms, some sex-specific, some not, controlled by different regulatory regions, with each version expressed in a different area of the growing fly.

Similarly, although it is convenient to think of the coding (THEN) region as encoding a single protein, nature is full of tricks for modulating (modifying) the shape of a single protein in different ways. For example, alternative splicing is a trick by which a single DNA sequence can be converted into several different but related proteins depending on the context; there have been suggestions that as many as half of the genes in the human genome are so modified.[18] A single fruit fly gene, *"Dscam,"* which is known to participate in the process of guiding axons to their destinations, can in principle be spliced into 38,016 different forms.[19] Humans have an equivalent gene, *"DSCAM,"* though we do not yet know whether it can be spliced in similar ways or what its functional significance might be.

But even if there are twice as many genes as we think, and each is on average spliced in a hundred different ways, we are left with something like 6 million proteins, far fewer than the number of cells in the body (trillions) or the number of cells in the brain (many billions). Some of that discrepancy might be brushed aside because, to the extent that the genome is a compression scheme, it is what computer scientists would call "lossy"—the output of the compression scheme is not identical to its input. JPEG compression, for example, does not perfectly preserve the image; the compression technique gets some of its considerable reduction by relying on the limitations of the human eye, and if you look carefully you can see that an uncompressed JPEG does not look exactly like its original. (Similarly, if you listen carefully, you can tell that an uncompressed MP3 doesn't sound exactly like the original.) Except in some very simple organisms, such as worms, two organisms that are "uncompressed" from the same genome will not be identical—evidence that the genome does not encode every detail of the final product. Cloned grasshoppers, for example, have slightly different neurons in slightly different configurations, and, as I mentioned in the opening chapter, the same holds for the brains (and bodies) of human twins.

And yet, we also saw that identical twins are pretty similar to each other, both in body and brain. If the genome is lossy, it's not *that* lossy; identical genomes do not guarantee perfectly identical bodies, but twins raised in the same womb tend to be born with quite similar brains.

I would point instead to four other factors at work in enabling a small number of genes to create the complexity of the human brain. First, as we have seen (and as the facts about lossiness suggest), the genome encodes structure not as a bitmap, but as a process. The CD-ROM's worth of information in the genome really wouldn't be enough to paint a bitmapped picture of an embryo, but it is enough to describe a process for building one. An artist who only wants to paint a picture that looks like a kind of tree has much less to remember than an artist who wants to paint a particular Ponderosa Pine from memory; in a similar way, if some alien's genome had to encode every cell in a body, it would need much more information (many more nucleotides) than our genomes do, because ours specify a general way to build a creature rather than an exact picture of every detail of the finished product. Our genomes are lossy because they specify methods rather than pictures, but it is precisely that lossiness that allows them to so efficiently supervise the construction of complex biological structure.

The second is that genes work in combination, not isolation. If there were a one-to-one correspondence between genes and cell types, with each new gene acting independently of all the rest, adding 100 genes to a genome could at most add 100 new kinds of cells. But as we saw at the end of Chapter 5, combinations of genes can work in tandem; 100 genes working on their own, for example, might simply specify 100 independent types of proteins, but, hearkening back to the chessboard analogy I made earlier, those 100 genes might break into two sets of 50 that when multiplied out could describe 2,500 new combinations. Because genes work in combination, the incremental effect of adding a new gene to a genome may be not linear, but exponential.

Third, genomes are what a computer scientist would call "extensible." Simple vector compression schemes have a limited number of "primitives" or "atoms" from which to compose their encodings;

they describe their pictures in terms of simple procedures for drawing basic elements such as squares, circles, and curves. The "compression scheme" of the body does the same thing, but with a twist. The genome's compression scheme can freely add new primitives; every gene can effectively serve as a new building block. The "master control gene" for eye formation, *Pax6,* for example, causes a fly to grow an eye on its antenna. One gene specifying the fate of millions of cells—a miraculous shorthand. Instead of requiring a separate set of instructions for every eye or for every rib, the genome can reuse the same master instructions, the same genes, over and over again, as often as necessary, simply by expressing them (switching them on) in multiple locations.

And this leads to the final reason why the "gene shortage" is really a nonproblem: Every gene (or nearly so) gets used many, many times. Rather than reserving a particular gene for each cell (which would require trillions of genes), the genome is set up to use nearly all of its individual genes in the development of many cells. Rather than assigning a new gene for each molecule of hemoglobin, the genome uses the same recipe over and over again. Centipedes and millipedes, for example, build many legs not by having a separate gene for each leg, but by expressing the same genes at the top of the leg-building cascades in many locations.

In both brain and body, gradients allow a special variation on this theme: A whole slew of cells can express the same gene, *but to different extents.* That one gene—one of those signal beacons distributed as a gradient—can serve as a guide to the entire group of cells, guiding each to slightly different destinations. Consider, for example, the set of neuronal connections known as "topographic maps" that run in parallel from the retina to the thalamus, much like the ribbon cables that used to connect a printer to a computer. To a first approximation, every cell in the retina has the same markers, but, in a variant of the gradually diffusing gradients mentioned in Chapter 5, the cells differ in how many of those markers (known as "Eph receptors") they have, with cells closer to the ear having more than cells closer to the nose. Individual growth cones use information about how many Eph receptors they have—and how many their neighbors have—to guide themselves to the appropriate target. Like schoolchildren lining up by height, the axons of individual ganglion cells sort themselves according to their Eph levels, moving to the head of the class or to the back depending on which axons have more. (The proof came when a team of biological all-stars got together to genetically alter a subset of mouse retinal ganglion cells, goosing their Eph levels by implanting a gene borrowed from chickens. Cells with elevated Eph receptors simply moved to the head of the class, forward in the midbrain, while their unaltered neighbors shifted back to make room.[20]) Such gradients allow thousands, even tens of thousands, of axons to organize themselves in a precise fashion using a tiny number or genes.

The beauty of such a system comes in its flexibility—the axons stemming from the retina can expand if there is more space in the midbrain, or fall in closer if there's less space than expected;[21] the same recipe can (like the line-up-by-height rule) be used for a dozen cells or a thousand. A brain built by pure blueprint would be at a loss if the slightest thing went wrong; a brain that is built by individual cells following self-regulating recipes has the freedom to adapt. There's no gene shortage, because nature has figured out how to use the same genes over and over—not as blueprints but as powerful, flexible recipes for constructing complex biological structure.[22]

SIM MIND

Showing that the genome has the wherewithal to guide the development of complex neural structure is not the same as showing that it can guide the development of complex *mental* structure. Complex mental structure presumably depends on complex neural structure, and other things being equal, organisms with more complex neural systems seem to have more complex cognitive systems. But a full understanding of the birth of the mind will ultimately rest on articulating the relationship between mental structure and neural structure.

It is fair to say that the field of cognitive neuroscience is not yet far enough along to have a "final theory" of the relationship, but a number of us are trying to move in this direction. In my own lab, we are taking tentative first steps toward such a theory by using computer simulations—to develop case studies of how new cognitive structures could evolve from old cognitive structures. In essence, what we're working toward is a kind of SimCity for the evolution of the mind, a computerized crucible for exploring how evolutionarily induced changes in developmental genes might lead to changes in cognitive structures.

For example, we have been studying how new cognitive functions could emerge from changes to the cascades for building the ribbon-cable-like topographic maps I described earlier. A wealth of research has shown that such maps, or things that look a lot like them, are spread throughout the brain, not just in the pathways that route visual sensation from the retina to the cortex (stopping in the thalamus), but also in those routing auditory sensation from the cochlea (the auditory version of the retina) to the cortex, those routing somatosensory input (such as touch and pressure information from the skin) to the cortex, and so forth. Topographic maps are also widely used within particular brain regions; for example, the visual cortex seems to consist of a large number of specialized areas, each of which seems to do a different kind of image-processing, a bit like a series of Photoshop filters that transform images by extracting edges or enhancing contrast.

Figure 8.1 Analyzing *Mona Lisa*

One of the central aims of the visual system is to separate figure from ground; before we can identify Mona as a person, or as a particular person, we have to separate her from the rest of the image. Programs such as Photoshop are filled with sophisticated tools for doing just that—"selection" tools like the MAGIC WAND, the MAGNETIC LASSO, and the GROW command—and the brain is, too. Studies of how people segment figure from ground show that they can use dozens of "cues," such as changes in color, brightness, contour, and so on, to decide what counts as figure and what counts as ground.

Figure 8.2   Segmented Mona

As a first test case, we have looked at one way that the brain might implement a version of Photoshop's GROW command. You use the GROW command by selecting part of the image and then asking Photoshop to highlight nearby parts of the image that look similar. You could, for example, select part of Mona's hands and then use GROW to select the rest.

The brain might well do something analogous, starting from a particular focus of attention and incrementally "looking around" to decide what else might belong to the same element of the image. Our interest is in how a neural circuit that implements the GROW command could itself grow, under genetic control.

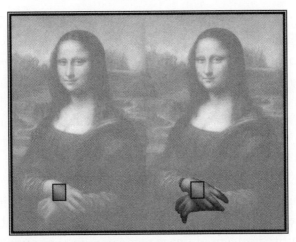

Figure 8.3   Segmenting Mona's hands by accretion

To see how this process might work, consider the logic of the GROW command. An image can be thought of as being made of thousands of pixels; the process of segmentation involves identifying which of those pixels belong to some coherent area within the image. For example, segmenting out Mona's hands requires the program to segment out the pixels in her hands. At each individual point in the image, a computer might ask, "Is this pixel of the right color (that is, the same shade of gray as other parts of the hand), and if it is, is it near enough to the other pixels that have already been highlighted?" A pixel in the face would fail the second test (being too far away), and a pixel in her robe would fail the first (being of the wrong color), but a nearby pixel in the hand would pass both tests and be added to the set of pixels defining Mona's hands. The set of pixels would gradually spread out in waves, with new pixels added each pass through, until no more pixels could be identified as belonging to Mona's hands, at which point the process of segmentation-through-accretion would be complete (although other neural processes might be required in order to complete the process of segmentation as a whole).

We have shown how this general idea of segmentation through accretion could be implemented in a "neural network"—a computer simulation of putative brain processes—consisting of three "layers" (or sets) of "neurons":[23] a set of "input" neurons standing for the retina, another set of "input" neurons standing for the focus of the network's attention, and a set of "output" neurons standing for the network's guess about how to segment the image.[24] Each of the output neurons handles a different part of the image and is wired such that it obeys exactly the logic described in the previous paragraph: "Is this pixel that I handle of the right color, and if it is, is it near enough to the other pixels that have already been highlighted?"

The net result—and you can either take my word for it, or turn to my web page (http://garymarcus.net) to see the simulation in action—is a relatively simple neural network that can solve a small part of a larger cognitive problem of image segmentation. The output "segmentation" layer does its work essentially by superimposing two topographic maps on top of one another—one from the retina to the segmentation layer, the other from the other input ("attention") layer to the segmentation layer—and by adding a few new neural connections so that each output neuron can obtain information as to whether its neighbors have been marked as part of the image.

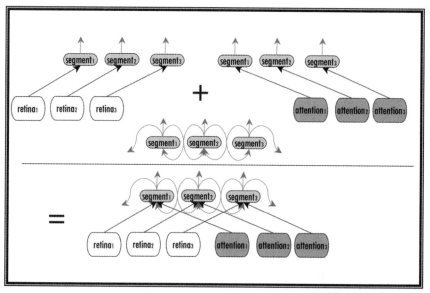

Figure 8.4   Segmentation through accretion, as a combination of simple maps

What's new and interesting here is not the mere fact that the network works, correctly segmenting very simple images—there are scores of other neural network simulations that explain how various simple aspects of cognition might work—but the way in which this particular network develops: not by a blueprint specified in advance (as is the usual practice in the field of neural networks), but according to a set of genes (which control processes such as the growth of axons and individual neurons). And not by a set of genes that originated from nowhere, but by slight modifications in the sets of genes already in place for the development of topographic maps. Although the details are complex, the take-home message is not. In essence, what we have shown is how GROW could be implemented in the brain *as a specialized variation of the more general circuitry* for building topographic maps, much in the way that the hand is a specialized variation of a more general system for building vertebrate limbs—a demonstration of how new cognitive systems might in principle have developed as variants of old ones—with the addition of only a few new genes.

9

FINAL FRONTIERS

Three stages of truth for scientists:
(1) It's not true.
(2) If it is true, it's not very important.
(3) We knew it all along.

—Leo Szilard

AT THE CORE of this book is a very simple idea, that what is good enough for the body is good enough for the brain, that the mechanisms that build brains are just extensions of those that build the body. Like Crick's "astonishing hypothesis"—the idea that the mind is a product of the brain—the idea that the brain is a product of the genes should be (to modern ears) scarcely surprising, an idea so natural we might wonder how we ever doubted it.

What's new here is the beginnings of a richer understanding of how genes work—alone, together, and in conjunction with the environment. In the twenty-first century, rather than thinking in terms of vague, almost undefinable abstractions such as nature and nurture, we will be able to understand the development of the mind and the brain in terms of specific biological mechanisms. Rather than trying to adjudicate "which one is better," we will try to understand how the two work together, as they do in the formation of ocular dominance columns, a perfect example of how the embryo seems to be endowed both with systems for creating structure independently of experience and with mechanisms for recalibrating those structures on the basis of experience. Ocular dominance columns in a mature organism are the product of both nature *and* nurture.

Genes and the environment truly are distinct: Genes provide options, and the environment (as well as the genes themselves, through their protein products) influences which options are taken. Blurring the distinction between the two—denying the dichotomy—would do little to aid in our understanding. But where does modern biology leave us in our understanding of the nature-nurture argument?

THE NATURE OF NATURE

Anyone who continues to doubt that genes play a significant, intricate role in shaping the mind is seriously mistaken. There is no gene shortage. Even a single new gene can, especially if it is at the top of a cascade, have an enormous effect. There is plenty of room in the genome to specify the initial structure of the brain in great detail. Yet our conceptions of "nature" and its contribution to the mind need significant reworking in the light of the gene.

The first problem with our traditional notion of nature's contribution to the mind is that it is far too static. Throughout popular culture, and even in the scientific literature, genes (or genomes) are most often treated as if they were straightforward portraits of our future from which we might read off our talents, our penchants, and our destinies. There is a powerful urge to think of genes as providing a single static contribution to the development of an organism, a primitive archetype or construction plan—what eighteenth-century German biologists called a *Bauplan*. In this view, something is innate if it is "specified" in the genome; if it is not specified in the genome, it is not innate. But, as we have seen, the relationship between genes and living beings is far too complex for that: Molecular biologists cannot simply discern from an organism's genome what its finished product will look like. The *Bicyclus anyana* butterfly (which, recall, grows up to be colorful if it is born in the rainy season but gray if it is born in the dry season) and fish that change their gender[1] (according to the presence or absence of a large, dominant male) show how obsolete the one genotype–one phenotype idea is. A single genome can be expressed in many different ways; there is no one-to-one mapping from genotype to phenotype. Indeed, even a single gene can be expressed in different ways, depending on which other genes are expressed around it and the signals it receives.

The second problem is that, although genes and the environment are truly distinct, any attempt to fully disentangle nature from nurture is doomed to failure. The actual realization of any genotype is always influenced by the embryonic environment. The clever conceit behind Michael Crichton's legendary *Jurassic Park* scenario—in which scientists could reconstruct a dinosaur from a preserved bit of prehistoric dinosaur DNA—glosses over the fact that even the earliest stages of gene expression are context-dependent: Every protein THEN has an IF, and from the moment of conception, many of those IFS are affected by the world that surrounds the growing embryo. Dino-DNA injected into frog eggs would likely yield something different from dino-DNA in dinosaur eggs—because the micro-environment of the egg would inevitably influence which genetic cascades were expressed. (Fans of the environment shouldn't get too comfortable, either—implanting a frog's DNA into a dinosaur egg would be even less likely to yield a dinosaur.) Because the recipes that build the mind and brain are always sensitive to the environment, there is no guarantee that those recipes will converge on any particular outcome, and there will never be an easy answer to our questions about nature and nurture.

The third problem is that we long for a simpler answer than we will find. If genomes were blueprints, understanding the origins of the mind would be simple. If we wanted to see whether some bit of neural structure was "innate," in the sense of being specified independently of experience, we could just look at the blueprint. If that bit of neural structure appeared in the blueprint, we could conclude that the structure was innate; otherwise, we would conclude that it was "learned." Deciding the nature-nurture controversy would be a simple matter of inventorying what was and was not spelled out in the blueprint, scarcely more complex than reading a map. But the straightforward, one-to-one mappings that we so strongly yearn for—from blueprint to brain, from brain to behavior—are not to be found; they might make for good engineering, but evolution hasn't built us that way.

Instead, in our world, nature's contribution to development comes not by providing a finely detailed sketch of a finished product, but by providing a complex system of self-regulating recipes. Those recipes provide for many different things—from the construction of enzymes and structural proteins to the construction of motors, transporters, receptors, and regulatory proteins—and thus there is no single, easily characterizable genetic contribution to the mind. In the ongoing, everyday functioning of the brain, genes supervise the construction of neurotransmitters, the metabolism of glucose, and the maintenance of synapses. In early development, they help to lay down a rough draft, guiding the specialization and migration of cells as well as the initial pattern of wiring. In synaptic strengthening, genes are a vital participant in a mechanism by which experience can alter the wiring of the brain (thereby influencing the way that an organism interprets and responds to the environment). There are at least as many different genetic contributions to the mind and brain as there are genes; each contributes by regulating a different process.

The final problem with the traditional notion of "nature" is that we tend to automatically equate it with "before birth," as if genes gave up their influence the moment the embryo left the factory. Even trained psychologists sometimes make this mistake, assuming that if a baby masters something early, the neural substrates for that something must be innate, or that something that happens late must be learned, ignoring the fact that something that is late could be as automatic as the sprouting of facial hair in a boy's adolescence. But genes are on board for a child's entire career, and on board through adulthood. We cannot exclude nature simply because something happens late in life (the symptoms of Huntington's disease, for example, only manifest themselves in adulthood, even though they can be reliably attributed to a gene that is inherited at conception), nor (as the Dr. Seuss-in-the-womb studies showed) can we exclude nurture simply because something happens in the womb. At a minimum, even in adults genes play a role in the consolidation of memory—the very route through which learning must pass—and it is perfectly possible that genes play more powerful roles in learning that we have not yet begun to understand. Genes aren't just for kids. Genes are for life.

In place of a view of the genome as a static blueprint that operates independently of experience and only up to the moment of birth, we have come to understand the genome as a complex, dynamic set of self-regulating recipes that actively modulate every step of life. Nature is not a dictator hell-bent on erecting the same building regardless of the environment, but a flexible Cub Scout prepared with contingency plans for many occasions.

THE NATURE OF NURTURE

Our notion of nurture, too, needs an overhaul. Just as there is no single contribution from the genes, there is no single contribution from the environment. Rather than relying on a single, monolithic, one-size-fits-all process of learning and information gathering, animals possess and make use of a diverse collection of neural and genetic adaptations for learning. Specialized neural circuits and genetic cascades allow organisms to extract specific bits of information from the environment and put them to particular uses. There is a wide range of tools in the genetic toolkit, from general ones enabling us to associate arbitrary bits of information to highly specialized ones providing us with a way of learning specific skills and information. For indigo buntings, this means a built-in tool for learning about the center of celestial rotation; for swamp sparrows, it means tools for being able to analyze and duplicate the structure of a song. And for *Homo sapiens,* it means tools for learning to communicate through language. Such mechanisms can be selectively impaired—an animal can habituate without being able to associate, or learn language while having trouble learning about the natural world (Williams syndrome) or vice versa (Specific Language Impairment).[2] There may be as many systems for learning as there are ways of detecting and analyzing information. What a creature can learn is a matter of what genes it has.

Without genes, learning would not exist. Genes support learning by guiding the growth of the neural structures that make learning possible and by participating (at least in some cases) in the very act of learning. In synaptic strengthening, for example, environmental signals (transduced by coincidence-detecting NMDA receptors) launch complex cascades of synapse-modifying genes. Such examples tell us that there can be no nurture without nature. They also suggest a more radical speculation: In principle, over the course of evolution, *any* genetic cascade may have come to be grist for the mill of environmental control—a speculation that suggests that traditional views of learning may have underestimated nurture as much as they have underestimated nature.

Just as one gene can participate in the construction of far more than a single neuron (provided other genes are already in place), nurture may be able to do far more with a single stimulus than change a single synapse. Although developmental psychologists often tend to think of learning as a slow, gradual process of synapse-by-synapse change, learning could—by tapping more broadly reaching cascades—trigger far more dramatic types of neuronal reorganization. For example, the learning that comes from experience in a rat's whisker could induce an entire new barrel field in the cortex, not by specifying each detail of the barrel field by some lengthy process of trial and error, but by recruiting cascades that have already evolved for constructing barrel fields independently of experience. Switching the social environment of a cichlid fish from one where it is submissive to one where it can dominate can trigger the expression of several dozen genes, leading to changes in color, increases in the sizes of some of the fish's neurons, and radical changes in its behavior. Such fish do not need to learn by trial and error how to behave as a dominant or submissive fish; instead, they use experience to switch between evolved genetic programs. Examples like these may just scratch the surface, giving only a tiny hint of how much an organism can make of experience when that experience is tied to complex cascades of gene expression.[3] Theories of what the environment could do for an organism have tended to float free of the genes, but real biological systems for exploiting the environment never do; where there is learning, there is underlying genetic mechanism, and where there is rich genetic regulation, there is the possibility of rich learning.

If there is no preformation, and no blueprint, there is also no getting away from the environment. Genes do not guarantee particular products; rather, they provide particular options: To every gene there is an IF, and with that IF comes an option. In many cases, those options are selected based on cues from the environment, and it is for that reason, more than any other, that the answer to the nature-nurture question is not one or the other, but both.

ALTERED STATES

Billions of years of blind tinkering has led to a system no engineer would have built from scratch—but that no biologist can fail to admire. If the system of self-assembly that nature has adopted is less straightforward than we might have hoped, it is also more elegant, anticipating some of the key ideas in modern computation—Boolean logic and parallel processing—by a billion years. Long before there was an Internet, eBay, or SETI@Home, the self-assembling toolkit of biology had found a way for thousands, billions, even trillions of cells to pass messages without relying on any central authority. Evolution has given us not a blueprint, but something like a self-organizing computer program of truly awe-inspiring complexity.

In the years to come, some of our best minds will try to dig deeper into that computer program, to figure out its individual lines of code (the IF-THENS that we call genes), the products of those lines (what we call proteins), how all those lines of biological code fit together, and how they make room for nurture.

In the long run, the effects on society will be profound. Take, for example, the advances that our increasing understanding of genes will lead to in medicine. Because, as we have seen, the brain is built like the rest of the body, it is also amenable to many of the same types of treatment. For example, stem cell therapies originally developed for leukemia are being adapted to treat Parkinson's disease and

Huntington's disease.[4] Gene therapies developed for cystic fibrosis may someday help treat brain tumors.[5] Both work by harnessing the body's own toolkit for development.

In stem cell therapy, physicians attempt to repair injured organs by injecting damaged tissue with special cells—stem cells—that have not yet completed the process of cellular specialization. Such cells can divide to create new cells and specialize to take on new tasks, picking up the slack from damaged cells. In essence, the goal of injecting the stem cells is to bootstrap the body into rebuilding itself, using as its guide the same genetic instructions it used during embryonic development. Gene therapy works (when it does—the technique still has a long way to go) by altering the body's complement of genes, supplementing (or replacing) aberrant genes with properly functioning counterparts. As scientists work out some of the rather substantial technical difficulties (for example, it is harder to inject genes into the brain than it is to inject them into other parts of the body, and harder to get the brain to take up those new genes), we can nevertheless expect stem cell transplants and gene therapy to be used in the treatment of both physical and mental disorders.

In the future, as gene therapy is combined with gene-splicing techniques, physicians may not just replace missing genes but add entirely new functions. For example, by splicing together customized IFS and THENS, physicians may be able to implant made-to-order genes that would enable the body to produce particular proteins as they are needed. Stanford biologist Robert Sapolsky has suggested that medical researchers may eventually be able to develop oxygen-sensitive genes that protect neurons from strokes and stress-sensitive genes that produce specific neurotransmitter receptors only when needed.[6]

In the meantime, genes have already quietly begun to play an important role in how doctors prescribe drugs.[7] Some oncologists decide whether or not to give their bone-marrow-transplant patients a particular drug based on the sequence of a particular gene that encodes an enzyme. Those with two copies of a gene that codes for a less efficient version of the enzyme are significantly more prone to side effects, and hence candidates for alternative treatments. Without genotype information, physicians might have had to abandon a drug that is effective for many patients because it was dangerous for a few. With continued progress in the rapidly growing field known as pharmacogenetics, it will become possible to prescribe drugs based on each patient's own unique biology.[8]

Over the next few decades, we will start to see mental disorders treated in the same way, with medications customized to particular genotypes.[9] For example, although Ritalin can be an effective treatment for people with some versions of Attention Deficit Disorder (ADD), it may be less effective with others. One preliminary study has suggested that such differences could be traced to differences in genes for dopamine receptors,[10] perhaps ultimately leading the way to genetic tests that could guide treatment. More generally, although complex mental disorders can rarely be traced to a single genetic anomaly, genetic techniques may help researchers track down which variants of a particular disorder are due to changes in enzymes, which are due to altered receptors, which are due to alterations in proteins that traffic in neurotransmitters, and so forth—knowledge that may ultimately allow doctors to customize treatment to individual genetic physiology. In the medicine of the twentieth century, brain and body were treated as largely separate; in the twenty-first century, the two will more and more come to be treated as the product of the same genetic toolkit.

From the fact that brain and body are assembled in the same way, it also follows that both can be altered using some of the same techniques. Scientists are already at work trying to build a bacterium with an artificially synthesized genome,[11] and it would be a small step from there to try to customize the primitive proto–mental life of that bacterium, to tweak that bacterium's complement of signals and receptors so that it could perceive and act in new ways. The gene-splicing tools that genetic engineers have used to create genetically engineered tomatoes can just as easily be used to modify the genes that underlie the brain. "Smart mice" are already on the drawing board (as we saw in Chapter 6), and although there are risks with any genetic modification, it may ultimately become possible to engineer cows that feel less pain or are less frightened of their handlers than normal cows, or dogs that are even more affectionate than a baby Golden. The final frontier, tampering with the human mind, cannot be far behind. Because there is no simple mapping between gene and function, "pure" genetic modifications may be rare; as we saw, tweaking NMDA receptors to make mice smarter also led them to experience more pain. It may be a long time before we can engineer babies that are even better at learning languages (or babies that don't lose their ability to learn language as they grow older), but it won't be that long before simpler traits that are shared by all mammals will, at least in principle, be easily modifiable. If it is possible to make a rat more affectionate by fiddling with its genes for vasopressin, there's a good chance that the same thing could be done with a human being.

In the coming decades, we will all—collectively, as a society—need to decide what we think about biotechnology and what applications we are and are not willing to allow. The debates we have now, about cloning and stem cell research, pale in comparison to debates we are likely to encounter as the technology for manipulating genes advances. We are already at the point where it is possible to screen embryos for their predisposition to certain life-threatening illnesses; as we unravel more and more of the genome, we will be able to detect more and more disorders (or predispositions to disorders) well in advance of birth. Ultimately, if we so choose, we may be able to directly manipulate embryonic genomes—add a gene here, delete a gene there. The genes of a child might eventually be more a matter of choice than chance.

With Gregory Stock, author of *Redesigning Humans,* and Francis Fukuyama, author of *Our Posthuman Future,* I believe that we will need to confront these choices sooner rather than later. In some cases, as Stock noted, the choices may be based more on fear than on rational argument. In the case of cloning, for example, much of the worry may be misplaced. What most people are afraid of is the creepy possibility that we might replicate a person, when in fact the most we could do is to replicate a *genome.*[12] As we have seen throughout this book, a genome is only part of the equation for building a person, an essential element but by no means the sole element. Even when two people do have identical genomes, they are still distinct people; to my mind, anyway, a clone is not so very different from an identical twin.

We should be far more worried about "genetic enhancement"—efforts to artificially construct "improved humans." Here I side with Fukuyama: Although the technology for improvement is close at hand, it comes with great risks, and some of the greatest risks stem from the complexity of the underlying biology.[13] As we have seen, the basic logic by which genes operate—the regulatory IF conjoined with protein template THEN—is straightforward—which is why genetic enhancement might be possible, in principle. But the combined effects of 30,000 genes far exceed our comprehension; if we know the general principles, we don't know the details, and what we don't know really could hurt us.

An analogy is with computer programs. Each individual instruction is perfectly understandable—there's nothing particularly mystifying or intrinsically unreliable about a line of a computer program that says, "If the contents of register A are greater than the contents of register B, execute subroutine C." But in a complex program, there's no reliable way to guarantee that registers A and B will always contain the right values, and at the wrong moment, subroutine C might well crash the program. Computer programs sometimes crash not because the individual lines of code are flawed, but because there is as yet no perfect way to foresee how they will all interact. Within the next few years, we will be able to synthesize whatever genes we like, splicing together a few favorite IFS and THENS, but it will take decades, maybe centuries, before we understand how those genes interact with one another. The upshot is that for many years it will be difficult, if not impossible, to gauge the potential side effects of a given manipulation in advance. I can live with a buggy beta-test version of a new software package, but I don't want to have to restart my child.

Not everybody is as risk-averse as I am; some day, maybe ten years from now, maybe a hundred, scientists will have gathered enough information about the genes that contribute to intelligence, athletic ability, and beauty, and enough sophistication with gene manipulation, that some adventuresome parents-to-be may think it is worth the risk.

At some point, our knowledge of how genes work might be so far beyond what we know today that the risk would be small, and the sort of person who willingly takes performance-enhancing drugs now (even knowing there are side effects) might well want to give Junior the best genome that money can buy. Whether that scenario becomes reality sooner or later, it behooves us to think deeply about it now.

CHANGING OUR DESTINY

Allow me to close with a few words about genes and inevitability. It is common to assume that if some trait—be it IQ or aggression, infidelity, or jealousy—is "genetic," it cannot be changed. Even as I look forward to knowing more about how genes work, I fear that as we gain that knowledge more and more people will feel that they have no control over their own lives, that we would be, in the words of British philosopher John Gray, "only currents in the drift of genes."[14] According to British science-writer Kenan Malik, "Many have read evolutionary accounts of human nature as explanations of human limits, as scientific validation of the impossibility of social solutions to our most deep-seated problems."[15]

And indeed, if we were built by blueprint and had no inborn mechanisms for rewiring our minds, "genetic" might well equate with "inevitability." But, precisely because genes are not blueprints but THENS that are always accompanied by IFS, there is no reason to read biology as leaving us no room for change. The idea that our genes might fix humans in a particular spot such that we cannot do differently is like the thought in the early 1950s that nobody could run a four-minute mile[16]—put to rest in May 1954, when Roger Bannister broke the barrier. Within forty-six days, someone else had, too, and by the end of 1957, sixteen runners had done it; by now it's a matter of routine. Just as our athletes are getting faster, our children may be getting smarter (though nobody knows why). A review of IQ tests over the century shows that scores have been steadily rising. On the British IQ tests known as Raven's Progressive matrices, almost everyone taking the test in 1967 (all but the bottom 5 percent) had higher scores than virtually everyone who took the test in 1877. In Holland (where some of the most reliable data have been collected), average IQ scores went up 20 points in the period from 1942 to 1982.[17] Throughout the world, infant mortality is down and literacy is up. Although progress is by no means inevitable, both societies and individuals can change, reaching targets that their predecessors thought were impossible.[18]

The new field of developmental cognitive neuroscience will, I hope, open new avenues by providing insight into the precise nature of the complex interactions between nature and nurture. To take just one example, consider a study published in the prestigious journal *Science* in 2002, one of the first to attempt to tie a specific mental trait to a specific gene-environment interaction. Psychologists Avshalom Caspi and Terrie Moffitt and their colleagues found that children with a certain version of a gene that produces an enzyme known as "MAO-A" (which metabolizes neurotransmitters such as serotonin and dopamine) are significantly more likely than other children to become violent—but *only* if they were mistreated as children—a case in which human behavior might be a bit like the body of the *B. anyana* butterfly.[19] Although the Caspi-Moffitt results establish only a correlation with the environment and not yet a causal relationship, there is good biological reason to find such results plausible. Many organisms (including humans) have a vast array of genes for dealing with stress, and the regulatory IFS that control the production of enzymes such as MAO-A might well be directly or indirectly sensitive to such stress.

Further studies of gene-environment interactions could eventually lead to a new way of identifying which children are at higher risk (be it for violence or for other social problems), and thus provide a new way of identifying children who might best profit from special day-care programs or home visits from social workers. Just as pharmacogenetics aims to match drugs to unique genetic physiology, a new field of therapeuto-genetics (for lack of a better word) could use individual genetics to prescribe customized social interventions. As we come to see genes not as rigid dictators of destiny but as rich providers of opportunity, we can begin to use our growing knowledge of nature as a means to make the most out of nurture.

APPENDIX:
METHODS FOR READING
THE GENOME

As ARCHITECT Mies van der Rohe famously said, "God is in the details." Each gene may be an autonomous agent that guides the synthesis of a particular protein, but what you get when you put them all together depends very much on the particular ones you've got.

The Human Genome Project—essentially a list (or "sequence") of the 3 billion DNA nucleotides that make up a human genome—is a first step toward understanding the details. But many of the headlines that followed the first successes in genome sequencing were seriously misleading, confusing the mere sequencing of the genome (that is, listing all the As, Cs, Gs, and Ts) with figuring out what all those nucleotides are for. A headline like "Researchers Decode Animal's Entire Genetic Blueprint"[1] makes it sound like we already know what all those ACGTs are for. But in truth, the problem is not just that the genome is not a blueprint, but that the 3-billion—long sequence of DNA nucleotides that make up our genome[2] is on its own nothing but a long list. According to a satire published in the literary magazine *McSweeney's*,[3] the "ten most surprising findings of the human genome project" were "GAGA," "CAAT," "GAGG," "AGGA," "TATT," "TACA," "GATT," "TAAC," "CATA," and "AAAA." The trick is to figure out what all those letters mean.

We do, of course, know enough to realize that the "words" in the language of DNA coding regions (THENS) have three letters (not four, but I forgive the *McSweeney's* editors their literary license). And we know how the coding region "words" get translated into protein sentences. But there is still a lot to learn about what those proteins are for and how they work together with their IF regions—especially when it comes to the mind.

Take schizophrenia as an example. Even though few people doubt that genes play a role—if one identical twin has schizophrenia, the chance that the other will have it is 45 percent, nearly three times the comparable number for nonidenticals[4] (and forty-five times the rate in the general population)—nobody really knows what role genes play in its development. John Nash, the Nobel Prize–winning economist depicted in Sylvia Nasar's 1998 book *A Beautiful Mind*,[5] and in the 2002 film of the same title, has schizophrenia, and so does his son, but scientists cannot yet point with confidence to any genetic "smoking gun." Scientists first claimed to have found a gene "for" schizophrenia in 1988,[6] but fifteen years and many quasi-discoveries later, the genetic basis of schizophrenia is still far from understood. A lot of the alleged genetic breakthroughs were probably simply false alarms; even a gene that does influence schizophrenia may be just one among many, and it may have different effects in different people depending on what other genes those people have and the environmental triggers they are exposed to.

Although scientists have had spectacular success in finding genes for single traits, there have been similar difficulties with autism, depression, alcoholism, and dozens of other disorders of the mind: Scientists are a lot better at "finding" genes for complex mental traits than they are at replicating their findings. It's like the old Mark Twain joke, "Quitting smoking is easy. I've done it a thousand times." As I'll explain below, it is somewhat easier to find a general region (locus) in which a gene is contained than it is to find that gene's precise location. When it comes to complex traits, gene-hunting will always be a challenge.

To understand relationships between specific genes and specific traits (or disorders), scientists use two main approaches: They work from known genes to figure out their functions, and they work backwards from known traits (or disorders) to find out which genes are involved. The latter approach was actually introduced first, almost ninety years ago, before Watson and Crick, and even before Avery, by a then nineteen-year-old college student, Alfred Sturtevant. Sturtevant, who worked in Thomas Hunt Morgan's famed Fly Room at Columbia University, began with an observation of Mendel's. Most of the time, if two different traits are tied to two different genes, they are "statistically independent" of one another, which is to say that having one tells you nothing about whether you will have the other. For example, knowing whether a pea is green or yellow tells you nothing about whether it will be smooth or wrinkled. Yellow peas were just as likely (no more, no less) as green peas to be wrinkled, short peas just as likely as tall peas to be smooth, and so forth, an observation that led Mendel to propose a law of "independent assortment."

But Sturtevant noticed that in some cases the law of independent assortment wasn't quite right. Sometimes two seemingly unrelated traits seemed to travel together. For example, Sturtevant's adviser, Morgan, had been trying to cross two kinds of fruit flies, a "wild type" (ordinary) brown fruit fly with normal wings, and an unusual fly with a black body and tiny "vestigial" wings. According to Mendel's law, you would expect body color and wing size to be independent, the chances of one separate from the chances of the other.[7] But, contrary to Mendel's prediction, the two traits, body color and wing size, *were* tied together—not because they were controlled by a single gene, but because they were controlled by two genes that were often inherited together. Because the traits for body color and wing size were "linked," black-bodied flies were much more likely than brown-bodied ones to have vestigial wings, and normal-winged flies were more likely than vestigial-winged flies to have brown bodies.

As Sturtevant studied other cases, he found that sometimes traits were completely independent (like two totally separate coin flips), sometimes they were tightly linked (like body color and wing size), and sometimes they were somewhere in between. Sturtevant's brilliant insight was this: Two traits that were influenced by two different genes would "travel together" only if those genes were on the same chromosome, and then, *the degree to which one trait could predict the other would depend on the physical proximity of those genes along that chromosome.* The closer together two genes were, the better one trait predicted the other. Using this observation, Sturtevant pieced together the first-ever gene map, a fairly accurate estimate of how six separate fly genes were ordered along a chromosome.[8] (Pretty amazing stuff considering that it was thirty years before Avery discovered that genes were made of DNA.)

What made all this work? It has to do with the process by which chromosomes are combined. As you'll recall, in almost every cell of your body, you have two copies of each chromosome, one from your mother, the other from your father. But rather than passing along both to your children, you pass along just a single, randomly selected copy of each chromosome. In the reduction of two chromosomes to one, known as *meiosis,* the maternal and paternal chromosomes get attached at various points and sections of the chromosomes are exchanged, crossing over to form a hybrid strand. If two genes are close together on a chromosome, they are likely to travel together, but if they are far apart, they are more likely to become separated by the process of crossing over. In this way, the probability of two genes traveling together provides a reliable estimate of genetic location.

For disorders that are controlled by a single gene, Sturtevant's calculus, known as "linkage," works amazingly well. Whereas he had only a handful of known anchor points to work with, modern scientists have huge libraries of reference points, known as *gene markers.* (Whether you realize it or not, you've probably already heard about them—markers form the basis not just of gene hunting but also of the now widespread technique of DNA fingerprinting.[9]) If a trait appears to be tied to a gene, but scientists don't know where that gene is, they can look at how often that trait appears in people who have various markers. If there is no link between having a given version of the marker and having that trait, they can safely guess that the gene they're looking for is on a different chromosome. But if there is a link between having the marker and having the trait, they can safely guess that the marker and the unknown gene are on the same chromosome, and with Sturtevant's math, they quickly zoom in on a rough estimate of the unknown gene's location.

Linkage can only take the gene hunter so far.[10] Crossover doesn't happen often enough to say *exactly* where to find the gene one is looking for; it only yields a rough neighborhood, getting within roughly one one-thousandth of the overall genome—fairly impressive but far from exact, enough to leave open a range of somewhere between a few hundred thousand and a few million nucleotides. For this reason, it is not uncommon to see reports saying that a particular trait has been tied, say, to a location on the short arm of chromosome 7, which contains hundreds of genes, rather than narrowing it down precisely to a single gene. This result would be disappointing if you wanted to know exactly which gene mattered, but it is impressive when you compare it with the 30,000 genes one starts out with. Even if linkage can't go that "final mile" by itself, it is a terrific technique for investigating the genetic basis of disorders that are caused by a single gene. The database known as Online Mendelian Inheritance in Man[11] catalogs thousands of genetic disorders, including many in which a single mutation, such as a deletion or an insertion, causes havoc. Cystic fibrosis and Duchenne Muscular Dystrophy are diseases of this sort, and linkage played a role in tracing the genes that were involved.

The bad news is that there are some circumstances where linkage doesn't work very well. Other gene-hunting techniques must be used (sometimes in association with linkage) for disorders that are influenced by many genes or disorders that can be caused by different genetic abnormalities in different individuals. Oversimplifying a bit, Sturtevant's math yields one number for one particular chromosomal location, and it does not gracefully handle situations in which there is more than one gene involved. The linkage method also runs into trouble when having a given gene doesn't guarantee that one will get the disorder, as when, for example, a disorder exhibits symptoms only in response to some relatively rare environmental trigger.

To deal with these more complex situations, geneticists have developed another class of strategies known as *association,* which typically involve directly correlating the nucleotides of the DNA with disorders that are under investigation. (Sturtevant's original linkage technique depended only on whether two traits co-occurred and did not make any reference to actual DNA nucleotides, which is why he could do it using fly family trees well before the era of Watson and Crick. Modern linkage analyses can use DNA nucleotides themselves as markers.) For example, researchers can test whether people with the sequence ACG-TAAT in the long arm of chromosome 17 are any more likely than people with the sequence TCGTAAT to have some mystery disease named *mysteremia.* Here, the genotype-phenotype connection needn't be perfect; if the chances were 30 percent versus 20 percent, the geneticists would still be in business. The trouble with association is that it's very prone to overinterpreting coincidence. On any given night, one might find that people who watched ABC were more likely than people who watched NBC to have heart attacks, but that wouldn't mean that watching ABC gives people heart attacks (memo to ABC lawyers—this example is purely hypothetical). It might be that on that night ABC drew more older viewers, or it might be sheer coincidence; it's highly unlikely that the higher number of heart attacks would be the *consequence* of the network's programming. An example from Eric Lander, the MIT scientist who led the public consortium's effort to map the human genome, helps make the point: "Suppose that a would-be geneticist set out to study the 'trait' of ability to eat with chopsticks in the San Francisco population by performing an association study with the HLA complex (a set of proteins important in the immune system). The allele HLA-A1 (a particular version of the HLA genes) would turn out to be positively associated with the ability to use chopsticks—not because immunological determinants play any role in manual dexterity but simply because the allele HLA-A1 is more common among Asians than Caucasians."[12]

Despite these problems, association tests are well worth doing. The problems can be avoided to some extent—by, for example, restricting studies to populations that are genetically homogeneous (for example, the population of countries such as Finland and Iceland that historically have been isolated)—and by combining them with Sturtevant's time-honored linkage techniques. The potential payoffs are huge—

when we do finally converge on a clear understanding of how genes are involved in complex disorders such as schizophrenia and depression, it's likely that association studies will have played an important role—and it's worth the wait. But we should be patient and not flinch when the technique occasionally leads to embarrassing failures.

UP FROM THE GENOME

Another strategy is to work in the opposite direction. Instead of starting with traits or disorders and hunting for responsible genes, biologists can start with candidate genes and try to work out their functions. Sequenced genomes, as we will see in a moment, provide some help in this process, but even with those sequences, the task of working from DNA to protein function turns out to be surprisingly difficult. For one thing, some DNA has no apparent function and may be little more than an accident of evolution, consisting of sequences that do not appear to correspond to proteins in modern humans.[13] (But much of what has been called "junk DNA" may turn out not to be; some so-called junk may contribute to gene regulation[14] or act as a "genetic scrapyard" from which evolution can scrounge bits and pieces.[15]) Even once nonfunctional DNA is excluded, it is not always obvious where the coding (protein template) region of a given gene begins, and even harder to figure out the boundaries of the regulatory IF part of a gene.

Next comes the even harder task of figuring out what a given protein is for. Putting aside the complications of processes such as alternative splicing (that trick for translating a single DNA sequence into several different but related proteins), it is a simple matter to use a sequence of DNA nucleotides to derive the correct sequence of amino acids in a given protein. But whereas the process of sequencing nucleotides is now essentially automated, there is no magic formula for figuring out a protein's function.

Traditionally, biologists have done so mainly by searching for mutants—flies with extra wings, mice with extra vertebrae, and so forth—a job that requires endless patience, in part because some mutations are so devastating that the poor animals who have them never make it through birth, so there is nothing to study. And among those organisms with altered genomes that do survive to birth, there is often no discernible effect at all, perhaps because some other protein has picked up the slack.[16] The monsters that get all the attention (and lead to the biggest scientific advances) are exceptions rather than the rule. Even where knocking out a gene (and its corresponding protein product) does have an effect, it is often subtle: Looking for variations can be as much an art as a science, tricky enough that NIH scientist Jacqueline Crawley wrote a book of advice for young biologists called *What's Wrong with My Mouse?*[17]

Despite these challenges, scientists have made enormous progress in understanding the function of individual proteins. One strategy, which has its origins in Edward Beadle and George Tatum's one-gene-one-enzyme irradiation experiments, is what one might think of as a "shotgun" approach: Scientists apply chemical and biological agents in an effort to mass-produce mutants. MIT biologist Nancy Hopkins, for example, has used retroviruses to create, essentially at random, hundreds of thousands of different zebrafish mutants, looking afterward for abnormal body structures, aberrant nervous systems, and unusual behaviors.[18]

Other researchers have followed a finely focused strategy, creating smaller numbers of more specifically targeted mutants, choosing a particular protein in advance, and exploring what happens in that protein's absence. All the knockout mice that we have seen are products of this strategy. Both strategies are necessary: Focused knockouts make it easy to assess the function of a particular gene, while randomly generated retroviral "screens" make it easier to get the big picture of what genes are involved in a given process.

Another approach to understanding a protein's function is to try to understand its shape. As we saw earlier, a protein is a complex three-dimensional structure, and its functions are governed in part by its twists and folds. So, it would be nice if scientists could directly take their pictures, but because the proteins are tiny, a thousand times smaller than the thickness of human hair, they must rely on much more indirect evidence. Techniques such as X-ray crystallography—which involves shooting X rays at laboriously prepared crystallized proteins—yield only achingly roundabout information, not straightforward images.

Luckily, most proteins are not sui generis; instead, they come in families, as a result of the way in which they evolved. Like human families, even though individual members differ, some traits are shared. In a human family, it might be the eyebrows, or a nose. In protein families, what is shared is short strings of amino acids known as *motifs.* For example, all the proteins in the "zinc finger" family contain a common stretch of amino acids that allows their host protein to bind to stretches of DNA or other proteins. These motifs simplify the process of working out a protein's structure, as well as its function. With state-of-the-art computer databases that are freely shared over the Internet, it has become easier and easier to work back and forth between the crystal structures and information about what is shared across proteins. The ultimate goal is to be able to create programs that take in DNA sequences as their input and produce models of folded proteins as their outputs. (Nature is way ahead of scientists on this one. Although biologists do their best to use the laws of chemistry and physics to build computerized simulations of the processes of protein folding, current simulations are only first approximations to the bending, twisting, and folding that actual proteins do spontaneously. Computers merely simulate the laws of physics; real amino acids have no choice but to follow them.)

Scientists today are equal-opportunity. To figure out the complex connections between genes, proteins, and function, they are prepared to use any technique, from linkage and association to X-ray crystallography, knockouts, and brain imaging. Because genes and proteins are so important to understanding both disorders and normal development, there is no doubt that it is worth the effort. But it will be a long haul. The next act after the Human Genome Project is the Human Proteome Project, an NIH-sponsored effort to catalog all the proteins in a human body.[19] The title of the first conference on the project says it all: "The Human Proteome Project: Genes Were Easy."[20]

FROM GENES TO BEHAVIOR?

Although the development of the brain in many ways parallels the development of the body, understanding the genetics of behavior poses special problems. For one thing, behavioral disorders, especially in contrast to physical attributes such as wing size or pea shape, are often difficult to diagnose—Mendel was lucky to be working with peas. Symptoms may vary from one day to the next because of mood or medication, and even when symptoms are stable it takes a lot more time to identify a mental disorder (or a behavioral abnormality in a mouse) than it does to measure a fly's wingspan. Furthermore, the knockout techniques that are widely used for studying the body do not necessarily depend on a mutant animal surviving past birth, but there's no way to study behavior in a mouse that doesn't make it.

There's an even worse problem that arises in studying any behavior that isn't pure reflex, and that's the way in which any one action is the process of many cognitive systems coming together. Think of the proverbial chicken, let's call her Henrietta, and why she might not have crossed the road. Henrietta might not have crossed the road because she couldn't see, because she couldn't make sense of what she saw (like a prosopagnosic, who can see but cannot recognize faces), because she was demoralized and didn't think that she could make it, because she didn't realize what was on the other side, or because (perhaps owing to a disease like Parkinson's) she couldn't translate her desire to cross into the right pattern of muscle activity. This ambiguity matters for studies of genes and behavior because almost everything we know about genes and behavior comes from studies of mutants that do not do some normal behavior; because most behaviors are the product of multiple subsystems, the mere absence of a behavior may mean many things.

The problem of studying the genetic bases of cognition and behavior is even more difficult when there are no direct "animal models" of a given trait. It's easy to tell when you have an albino mouse, but more difficult to tell whether an "anxious" mouse is truly anxious, and even harder to figure out what a reasonable mouse model of specific language impairment might be.

Until just a few years ago, many biologists didn't seem to fully appreciate these points. If a mouse didn't find its way through a maze that it had seen before (a common measure of memory), biologists would assume that their mouse failed because it had trouble remembering where the food was. Now, most biologists have come to realize what psychologists could have told them all along: A so-called "failure" might have nothing to do with memory. A mouse might fail because it was stressed out (the leading memory test, for example, known as the "water maze," forces terrestrial rats to swim, not something that rodents are particularly fond of), because it couldn't see, or because, like some mice, when given a choice, it prefers to sink rather than to swim. No failure (and this should remind you of Chomsky's point about competence and performance) is ever easy to interpret. Even when a transgenic mouse does succeed, experimenters have to be careful that the test animal really is relying on memory and not, say, some smell that is too subtle for the experimenter to notice. Most interesting behavior is the product of a complex interplay between a wide variety of neural systems for perception, attention, decision making, motivation, and more, and so the mapping between genes and behavior is never simple.

Further progress will depend, more than anything else, on cooperation between scientists from a wide range of disciplines. Every aspect of the technology is getting better, from the techniques for knocking out individual genes to freely distributed computer programs such as BLAST[21] (Basic Local Alignment Search Tool), which allows biologists to get quick answers to questions like "Does *FOXP2* resemble any gene in the fruit fly genome?" There are also now databases of the three-dimensional structures that allow researchers to more quickly understand the physical differences between related proteins.[22] But even more important will be collaboration. In the quest to unravel the complex interaction between behavior, brain, and genome, there can be no substitute: Geneticists, molecular biologists, neuroscientists, psychologists, linguists, and even chemists and physicists, will all need to work together.

GLOSSARY

activity, brain. Neuronally induced electrical events in the brain; measured indirectly in brain-scanning techniques such as Functional Magnetic Resonance Imaging (fMRI) and Positron Emission Tomography (PET).

adaptation. A feature or a trait that has evolved because of its selective advantage, such as the dark gray color of moths during the Industrial Revolution.

adrenaline. A hormone that stimulates blood pressure and heartbeat.

alkaptonuria. One of Simon Garrod's inborn errors of metabolism, a disorder that darkens the urine of its victims.

alternative splicing. A process by which a single strand of coding DNA may be translated into one of several different strands of RNA, each in turn leading to the assembly of a different protein. See also *DNA, RNA.*

amino acid. Any of the twenty or so different organic acids, such as valine and serine, from which proteins are composed. See also *protein.*

amphioxus. Small, flat, jawless fishlike organisms thought to be closely related to the first vertebrates.

amygdala. An almond-shaped brain structure in the temporal lobe; involved in emotional responses.

amyotropic lateral sclerosis (Lou Gehrig's disease). A progressive, neurogenerative disorder that affects motor neurons and hence motor control.

animal model. A laboratory animal used to study another, typically more complex or harder to study organism, as when, for example, scientists study the effects of smoking on laboratory mice in order to estimate comparable effects on humans. See also *model organism.*

association (in genetics). A technique for studying whether particular genes (or genetic markers) are correlated with particular traits. See also *genes, genetic markers.*

association (in psychology). A link between a stimulus (e.g., a bell) and a response (e.g., food). See also *associative learning.*

associative learning. A process of learning relations between a stimulus and its response based on their co-occurrence in time.

attention. The neurocognitive systems that select what aspects of the environment (or which internal cognitive states) merit further analysis.

auditory area. Part of the temporal lobe devoted to processing sound.

autism. A disorder that affects social cognition.

autonomous agent. A metaphor for genes that emphasizes the fact that each individual gene is expressed according to its own unique set of regulatory IF conditions. See also *IF-THEN.*

axon. The long, spindly output part of a neuron that carries electrical impulses from the body of the neuron to its target. See also *axon guidance molecules, dendrite, myelin, neuron.*

axon guidance molecules. Chemical cues that guide developing (or regenerating) axons to their proper destinations. See also *axon, cell adhesion molecules, growth cones, neurotrophins, Robo.*

azimuth system. A system (in, e.g., honeybees) for determining an animal's orientation based on the angle of the sun on the horizon.

barrel fields. In rodents, areas of sensory cortex that respond to whisker stimulation.

basal ganglia. A subcortical group of neurons found at the base of the brain; traditionally implicated in motor control, and more recently thought to also be important for language.

base, DNA. See *nucleotide.*

behavioral genetics. A scientific approach that seeks to understand the genetic and environmental contributions to behavior using techniques such as the comparison of identical and fraternal twins.

blueprint. As used here, an obsolete metaphor that would identify genomes with detailed pictures or schematics of finished products.

C. elegans (*Caenorhabditis elegans*). A translucent roundworm about 1 millimeter in length that serves as one of the main model organisms in modern biology. See also *model organisms.*

cadherins. Cell adhesion molecules that guide processes such as cell migration and axon guidance. See also *axon guidance molecules, cell adhesion molecules.*

CaM Kinase II. An enzyme involved in energy transfer that appears to play an important role in synaptic strengthening. See also *synaptic strengthening.*

cascade. A collection of linked genes, such as the group of eye-formation genes driven by *Pax6.* See also *master control gene, Pax6.*

cell. A basic unit of biological structure, typically consisting of a nucleus, a membrane, and a variety of specialized organelles. See also *cell membrane; cytoskeleton; death, (planned) cell; differentiation, cell; division, cell; neuron; organelle.*

cell adhesion molecules. Sticky molecules that help guide neurons and their processes (axons and dendrites) to their destinations; also used in the migration of other types of cells. See also *axon, axon guidance molecules, dendrite.*

cell membrane. A thin layer of fat and protein that separates the internal environment of a cell from its external environment.

cerebrum. The largest part of the forebrain, and indeed of the brain itself; divided into two hemispheres. See also *cortex, cerebral; cortex, prefrontal; forebrain, hemisphere, left; hemisphere, right; neocortex; somatosensory cortex.*

channels. See *ion channels.*

chromosome. Coiled groups of DNA strands found in a cell's nucleus.

clone. An organism (or cell) that is formed from a copy of another organism's (or cell's) DNA.

coding (THEN) region. The part of a gene that specifies a protein template. See also *IF-THEN, regulatory (IF) region.*

combinatorial cues (in gene expression). A system by which the regulation of a given gene depends on the presence (or absence) of multiple molecules.

competence. Noam Chomsky's term for a person's underlying or ultimate knowledge of grammar; used here to refer to the abstract ability to do something, as contrasted with a person's actual performance. See also *performance.*

compression. In computer science, techniques for reducing the amount of information required to store a particular file; as used here, a perspective on how a small genome might relate to a complex phenotype.

congenital. Present from birth (but not necessarily genetic).

convolution. As used here, one of the folds of the cortex. See also *cerebrum; cortex, cerebral.*

corpus callosum. A thick tract of neural fibers that serves as the major connection between the left and right hemispheres of the cerebrum; when severed, split-brain syndrome results. See also *cerebrum; hemisphere, left; hemisphere, right.*

cortex, cerebral. The outer surface of the cerebrum (itself the major part of the forebrain); essential for language and higher-level cognitive function. In mammals, consists both of paleocortex (including, for example, the limbic system) and the neocortex. See also *cerebrum; cortex, prefrontal; forebrain; neocortex; somatosensory cortex.*

cortex, prefrontal. The frontmost portion of the cortex, implicated in reasoning, decision making, and emotion. See also *cerebrum; cortex, cerebral; forebrain; neocortex; somatosensory cortex.*

CpG island. A short region of DNA with a high proportion of repeated CG nucleotides, often associated with the regulatory region of a gene.

cystic fibrosis. A disease affecting the lungs and pancreas often caused by a single genetic error in a protein involved in the traffic of sodium and chloride ions.

cytoskeleton. A network of filaments that provides structural support to a cell.

death, (planned) cell. Also known as *apoptosis,* the process by which cells are induced to die as a programmed part of the normal developmental process, such as in the loss of webbing between embryonic fingers. See also *cell.*

dendrite. The input part of a neuron; carries input from the synapse into the cell body.

differentiation, cell. The process by which cells take on specialized identities. See also *cell.*

dishabituation. The process by which an organism that has grown progressively less sensitive to a particular repeated stimulus grows to again be sensitive to that stimulus. See also *habituation.*

division, cell. The process by which a cell gives rise to progeny; may be either symmetric (yielding two similar or identical progeny), or asymmetric (giving rise to two nonidentical progeny, such as when a stem cell gives rise to an uncommitted stem cell much like itself, and a more differentiated daughter cell). See also *cell.*

DNA (deoxyribonucleic acid). The double-helix-shaped molecule that serves as the principal repository of genetic information; made up of two sugar-phosphate backbones held together by pairs of DNA nucleotides. See also *nucleotide, RNA*.

DNA hybridization. A process for combining or comparing two pieces of single-stranded DNA. See also *DNA*.

duplication, gene. A mutational process by which the DNA sequences corresponding to genes (or even whole chromosomes) are inadvertently repeated in a copied gene or chromosome; a major factor in evolutionary change.

dyslexia. A severe disorder of reading not attributable to other cognitive, memory, or motivational impairments.

E. coli *(Escherichia coli).* An intestinal bacterium that has been intensively studied as a model genetic organism. See also *model organism*.

early-response genes. A specific set of genes that are expressed quickly in response to specific kinds of stimulation, ultimately leading to the triggering of other genes.

electrical impulse. The all-or-nothing electrical signal transmitted by a neuron; also known as an action potential.

elegans. See *C. elegans*.

ELH (Egg-laying hormone). A gene found in the sea slug *Aplysia* that yields a protein that plays several roles in triggering the complex steps of its egg-laying behavior.

embryo. As used here, an organism from conception to birth (or hatching); also sometimes used to refer specifically to the earlier stages in development, as in a human offspring from conception through the eighth week (at which time the offspring may be referred to as a *fetus*).

empiricism. As used here, the belief that little beyond the apparatus of sensation (sight, smell, and so forth) is prewired into the brain and mind. See also *nativism*.

Emx. A family of genes important in establishing the identity of particular brain regions.

enzyme. A protein (or strand of RNA, or protein complex that consists of a protein and other factors such as sugars) that serves as a catalyst to promote a specific biochemical reaction. See also *protein*.

epigenesis. As used here, a process of successive approximation; also used (elsewhere) to refer to a process that supplements or modifies gene expression. See also *successive approximation*.

epithelial cell. A type of cell that typically serves as part of a thin layer of protective lining in various organs such as the lung and guts; also the outer surface of the body.

equipotentiality. The apparently incorrect idealization that any brain cell or region could take on any function.

expression, gene. See *gene expression.*

FGF8 (fibroblast growth factor 8). A gene whose versatile protein product plays a critical role in establishing the basic geography of the cortex.

filopodia. Fingerlike tips of growth cones that explore the extracellular environment. See also *growth cones.*

forebrain. Frontward part of the brain that includes the cerebral hemispheres and structures such as the thalamus and hypothalamus. See also *cerebrum, midbrain, hindbrain.*

FOXP2. The first gene to be directly implicated in a speech and language disorder; found in many species, but notably different in humans.

Fragile X syndrome. A common form of genetically inherited mental retardation that has been linked to abnormal patterns of nucleotide repetition in a specific gene *(FMR1)* on the X chromosome. See also *X chromosome.*

Fru. A gene that appears to contribute in several ways to fruit-fly courtship.

Functional Magnetic Resonance Imaging (fMRI). A brain imaging technique that traces local changes in blood flow as an index to brain activity. See also *activity, brain; Magnetic Resonance Imaging (MRI).*

GABA. An inhibitory neurotransmitter that carries messages between neurons. See also *neurotransmission, neurotransmitter.*

gene. A combination of IF and THEN that specifies a protein template and conditions for that protein's production, such as the genetic recipe for insulin and coupled instructions that govern where (i.e., in the pancreas) that recipe should be followed. See also *gene expression, genetic markers, genome, IF-THEN.*

gene expression. The process by which a gene is converted into protein (after being transcribed into RNA). See also *protein; RNA; transcription, gene.*

gene shortage. Paul Ehrlich's term for the discrepancy between the relatively small number of genes and the relatively large number of neurons.

gene transcription. See *transcription, gene.*

genetic markers. Particular identifiable sequences of DNA with known locations on a chromosome that can be used in helping to identify the location of unknown genes.

genome. The total collection of a given organism's genes.

genotype. A given organism's genome, contrasted with the organism's appearance and behaviors, or phenotype. See also *phenotype.*

glia (or glial cell). Neural support cell that produces oligodendrocytes (responsible for the myelin sheath that insulates neurons) and astrocytes (which produce mechanical and metabolic support); such cells are also implicated in guiding neural migration.

glutamate. An excitatory neurotransmitter that carries messages between neurons. See also *neurotransmission, neurotransmitter.*

gradient. A molecular marker that changes in concentration over space, providing spatial cues to growing cells; examples in the text include *Emx,* which contributes to the development and specialization of cortical areas, and *Eph,* which contributes to the development of topographic maps. See also *Emx, topographic map.*

growth cones. The migrating tip of a growing axon, which helps to guide that axon to its destination. See also *axon, axon guidance molecules.*

habituation. The process by which an organism learns to ignore (gets used to) a particular ongoing stimulus; also, an experimental method with which to test the perceptual or cognitive abilities of an organism such as a human infant. See also *association; dishabituation; learning mechanism, general.*

hardwired. Fixed in advance and not modifiable. See also *prewired.*

Head Start. An early intervention program designed in 1965 to provide at-risk preschool (0-5) children with educational, social, nutritional, and medical assistance.

hemisphere, left. The left half of the cerebrum. Through connections that pass through the corpus callosum, controls the right half of the body; often associated with language and analytic thinking. See also *cerebrum; corpus callosum; hemisphere, right.*

hemisphere, right. The right half of the cerebrum. Through connections that pass through the corpus callosum, controls the left half of the body. See also *cerebrum; corpus callosum; hemisphere, left.*

hemoglobin. The iron-bearing protein in red blood cells that carries oxygen from the lungs to the body's tissues.

heritability. A measure of the correlation between individual variation in a trait and genetic relatedness.

hierarchical tree structure. See *tree structure.*

hindbrain. Rearmost part of the brain; includes pons, medulla, and cerebellum. See also *forebrain, midbrain.*

hippocampus. Cortical structure in the temporal lobe thought to play an important role in short-term memory and spatial representation.

homeobox genes. A broad family (or "superfamily") of evolutionary related regulatory genes, including the *Otx* family, the *Pax* family, and the *Hox* genes that control the expression of many other genes. See also *Otx, homeodomain, Hox, Pax6.*

homeodomain. A particular stretch of DNA, corresponding to approximately sixty amino acids, found in homeobox genes; binds to the DNA of other genes, influencing their regulation. See also *homeobox genes.*

Hox. A particular family of homeobox genes that are important in establishing basic body plans through the control of cell differentiation; found in animals from fruit flies to humans. See also *homeobox genes; differentiation, cell.*

Human Genome Project. The late-twentieth-century project to determine or "sequence" all the nucleotides in a human genome, now largely complete. See also *genome, nucleotide.*

Huntington's disease. An adult-onset disorder associated with jerky, involuntary movements, emotional disturbance, and loss of intellectual function as the result of progressive degeneration of neurons; tied to excessive repeats of the nucleotide sequence CAG in the Huntingtin gene.

hybridization, DNA. See *DNA hybridization.*

IF-THEN. The combination of *regulatory region* (IF) and *coding region* (THEN) that determines the protein product of a gene and the circumstances under which that protein should be synthesized. See also *coding (THEN) region; regulatory (IF) region.*

imaging, brain. Techniques such as Functional Magnetic Resonance Imaging (fMRI) and Positron Emission Tomography (PET) used to visualize the structure and function of the brains of organisms. See also *activity, brain.*

individual differences. Differences between members of a species, such as from one person to the next in height, weight, or IQ.

innateness. See *nativism.*

input. Information that feeds into a brain area or computation. See also *output.*

instinct. A prewired or hardwired neurocognitive system. See also *hardwired, prewired.*

insulin. A pancreatic hormone that regulates the level of glucose in the blood.

invertebrate. Animals such as insects and worms that lack a backbone. See also *vertebrate.*

ion. A positively or negatively charged atom.

ion channels. Protein pores through which ions can flow. See also *ion.*

junk DNA. DNA with no known function. Some so-called junk DNA may be a relic of DNA sequences that had a purpose earlier in evolution; other junk DNA may simply be DNA for which the contemporary function has not yet been unraveled.

Kallman syndrome. A congenital disorder of the hypothalamus that affects smell and sexual development.

knockout mouse. A mouse that has been genetically engineered such that a particular gene is rendered permanently inactive. See also *transgenic animal.*

language acquisition. The process of learning a language.

language acquisition device (or language instinct). Hypothesized inborn machinery for learning language.

learning mechanism, general. A neurocognitive device for learning that is general-purpose, not specialized for the acquisition of any particular kind of information. See also *learning mechanism, specialized.*

learning mechanism, specialized. A neurocognitive device for learning that is tuned for the acquisition of particular types of information, such as language or social relations. See also *learning mechanism, general.*

lesion, brain. A localized injury to the brain.

linkage. A technique for finding the location of an unknown gene based on its pattern of co-occurrence with known genes.

lissencephaly. "Smooth brain," a disorder resulting in mental retardation in which the brain lacks its usual convolutions. See also *convolution.*

locus. A region of a chromosome with which a particular trait or disorder has been identified.

long-term memory. The cognitive system that stores memories indefinitely, for months or even years. See also *short-term memory*.

long-term potentiation (LTP). See *synaptic strengthening*.

Lou Gehrig's disease. See *amyotropic lateral sclerosis*.

LTP. See *synaptic strengthening*.

Magnetic Resonance Imaging (MRI). A technique for generating pictures of the brain or body of an organism. See also *Functional Magnetic Resonance Imaging (fMRI)*.

master control gene. A gene such as *Pax6* that is atop a genetic cascade. See also *cascade*.

matter, gray. The part of the brain and spinal cord made up mainly of neuronal cell bodies. See also *matter, white*.

matter, white. The part of the brain that is largely made up of myelinated nerve fibers. See also *matter, gray*.

Mauthner neurons. Neurons involved in escape and startle reflexes that are unusually large (and hence unusually accessible for experimentation); found in fish and closely related species.

medial geniculate nucleus. A way station in the thalamus that transfers auditory information from the cochlea to the cortex. See also *thalamus*.

membrane. See *cell membrane*.

mental life. The cognitive activities of an organism, including its thoughts, beliefs, desires, intentions, and goals.

metabolism. Chemical interactions in a living organism that are essential for life, either breaking down substances to release energy and nutrients or synthesizing substances.

midbrain. Central part of the brain; coordinates visual and auditory reflexes and controls functions such as eye movements. See also *forebrain, hindbrain*.

midline. Imaginary plane that divides the left and right hemispheres of the cerebrum or the left and right halves of the body. See also *cerebrum; hemisphere, left; hemisphere, right*.

migration, cell. A process in which cells move from their origination points to their final destinations.

mitochondria (plural of *mitochondrion*). Energy-generating powerplants of the cell.

model organism. Intensively studied organisms such as yeast, *C. elegans* roundworms, *D. melanogaster* fruit flies, zebra fish, and mice; chosen to provide insight into how particular biological systems work in other, more difficult to work with organisms (such as humans). See also *animal model.*

modularity hypothesis. The hypothesis that significant portions of the neural or cognitive system consist of circuits (or brain areas) dedicated to specific functions.

module (or mental module). A region or circuit of the brain that is specialized for a particular function.

motor control. The cognitive system responsible for the coordination and control of muscles (or other effectors).

motor neurons. Neurons that drive effector cells such as muscles.

muscular dystrophy. A disorder in which there is a gradual deterioration of skeletal muscles; at least one version has been tied to deletions in a particular gene.

mutation. A random change in the genetic code, as the result of processes such as miscopying; may be harmful, beneficial, or neutral.

myelin. A whitish insulating material made of proteins and fat that surrounds many of the axons in most vertebrates. See also *axon.*

nativism. The belief that some complex cognitive and perceptual structures are prewired. See also *empiricism.*

nematode. See *C. elegans.*

neocortex. The six-layered outer sheet of the mammalian cerebral cortex. See also *cerebrum; cortex, cerebral.*

nerve cell. See *neuron.*

nerve net. A diffuse set of interconnected neurons, such as those found in a jellyfish, that lacks the structure and centralization of vertebrate central nervous systems.

nervous system, central. The brain and spinal cord.

neurofibromatosis. A genetically linked learning disorder.

neuron. A cell specialized for computation and communication; the basic building block of the brain and spinal cord.

neurotransmission. The process by which neurons communicate, involving the release of neurotransmitters across a synapse. See also *neurotransmitter.*

neurotransmitter. A chemical messenger, such as serotonin or dopamine, that carries a signal between different nerve cells. See also *neurotransmission, receptor.*

neurotrophins. Long-range signaling molecules that guide axons and influence neuronal cell survival. See also *axon, axon guidance molecules.*

NMDA receptor. A particular kind of glutamate receptor that appears to play a critical role in learning and memory; may serve as a "coincidence detector" that allows calcium ions to flow only (or maximally) when the receipt of a glutamate signal coincides with a particular electrical change (depolarization). See also *glutamate, ion, receptor.*

nucleotide. An individual "letter" of DNA consisting of a base (adenine, guanine, thymine, or cytosine), a molecule of sugar, and a molecule of phosphoric acid. See also *DNA.*

object permanence. The belief that objects continue to exist even when they are removed from view.

object recognition. The neurocognitive ability to recognize objects based on patterns of light cast on the retina.

ocular dominance columns (or ocular dominance stripes). Pattern of alternating stripes of neurons in the cortex; neurons in a particular stripe are driven largely or exclusively by a single eye.

olfactory system. A system for detecting smells.

one-to-one correspondence. A mapping between two systems in which every input has a unique corresponding output, such as the mathematical function $f(x) = x$, or a topographic map in which a location in the retina connects to a unique corresponding location in the tectum. See also *tectum, topographic map.*

optic nerve. A bundle of nerve fibers that transmit information from the eye to the brain.

organelle. Specialized cellular substructures such as mitochondria, nucleus, and the endoplasmic reticulum.

orientation map. Neural circuits that are sensitive to the orientation of lines.

Otx. A family of homeodomain-containing genes important in brain development. See also *homeodomain.*

output. The product of a (neural) computation. See also *input*.

Parkinson's disease. A progressive disorder affecting motor control caused largely by the progressive degeneration of dopamine-producing neurons in a specific site in the midbrain.

pathway. A set of neural connections from one brain area to another.

Pax6. A homeodomain-containing gene that plays an important role in eye development and brain formation. See also *homeodomain*.

performance. Noam Chomsky's term for in-practice factors such as memory or attention that may limit the ability of some system to fully express its underlying knowledge, or competence. See also *competence*.

phenotype. The appearance and behaviors of an organism, in contrast to its genotype. See also *genotype*.

phenylketonuria (PKU). A version of mental retardation that stems from the inability to metabolize the amino acid phenylalanine.

pituitary gland. A pea-sized, hormone-producing gland at the base of the brain.

PKU. See *phenylketonuria*.

planum temporale. The upper surface of the temporal lobe thought to be involved in language processing; typically larger in the left side of the brain than in the right.

plasticity. The capability of the brain to change in response to experience or damage.

preformationism. The theory, popular in the seventeenth century, that a sperm cell or egg cell contained a miniature embryo.

presumptive. A precursor cell or tissue that under ordinary circumstances is fated to develop a particular way; for example, a presumptive eye cell.

prewired. As used here, a brain structure that develops in advance of experience that may or may not be modifiable in later development. See also *hardwired*.

protein. One of the basic building blocks of a cell, consisting of a string of amino acids bent or folded into three-dimensional structures that support complex functions; examples include insulin, collagen, and transcription factors such as *Pax6*. See also *amino acid, enzyme, protein synthesis, protein template, regulatory protein*.

protein synthesis. The process of converting from DNA to RNA to protein.

protein template. The coding region of a gene that dictates what protein will be synthesized if that gene is expressed. See also *coding (THEN) region, IF-THEN.*

protocadherin. A special type of cell adhesion molecule that may play an important role in the development of the brain. See also *cell adhesion molecules.*

receptor. Protein sentinels that span the cell's membrane, receiving extracellular signals on the outside and conveying news of their receipt to the inside of the cell. Each receptor is specialized to receive a particular signal such as a hormone or neurotransmitter.

recursion. In computer science, the ability of a program or procedure to call itself. As used here, a process by which progressively more complex elements (such as linguistic sentences) can be assembled out of combinations of simpler elements; these complex elements can then serve as the input to still more complex elements.

reelin. A protein product apparently found only in vertebrates that appears to contribute to both axon branching and the process of synapse generation; named after *reeler,* a mutant mouse that wobbles around as though drunk.

reflex. An automatic reaction in which a perceptual input leads immediately to an action, such as the patellar (knee-jerk) reflex, or the rooting reflex of an infant.

regulation, gene. The process that controls the expression of a given gene. See also *gene expression.*

regulatory (IF) region. The part of a gene that specifies the conditions for that gene's expression. See also *gene expression, IF-THEN, coding (THEN) region.*

regulatory protein. A protein that controls the regulation of other genes. See also *regulation, gene.*

representational format. The way in which information is arranged and encoded in memory.

retina. The part of the eye that transduces light into electrical activity.

retinoic acid. A growth factor important in limb development and (in organisms such as salamanders) limb regeneration; may also play a role in neural regeneration.

RNA (ribonucleic acid). The complement of DNA that serves as the intermediary molecular template in the process of protein synthesis. See also *DNA.*

Robo. A family of genes implicated in the guidance of axons across the midline. See also *axon, midline.*

schizophrenia. A disorder of impaired thinking, often including symptoms such as hallucinations, paranoia, and social withdrawal.

serotonin. A neurotransmitter implicated in the regulation of mood, appetite, and arousal.

short-term memory. A neurocognitive system that stores memories for seconds or minutes. See also *long-term memory.*

sickle-cell anemia. A disorder in which blood cells, having formed a sickled shape, stack together, impairing circulation; tied to a specific single-nucleotide change in a gene for hemoglobin. See also *hemoglobin, nucleotide.*

social cognition. The neurocognitive systems that are responsible for understanding, predicting, and guiding interactions between individuals.

somatosensory cortex. The part of the cortex dedicated to the processing of the sensations of touch, pressure, temperature, and pain.

Specific Language Impairment. A disorder of language not attributable to other cognitive, memory, or auditory impairments.

split-brain syndrome. Syndrome in which the major conduit of communication between left and right hemispheres is severed. See also *corpus callosum.*

statistical information. Quantitative information that describes the properties of entities, such as how tightly two things are correlated or how frequently a particular feature occurs in a population.

stem cell. A precursor cell that can reproduce itself and give rise to a variety of different types of progeny cells.

stereotyped behavior. A complex behavior that proceeds in a stylized, unvarying way.

stimulus. An input or cue from the environment that animals may analyze and use as a cue to further action.

striatum. Part of the basal ganglia. See also *basal ganglia.*

successive approximation. A process of gradual refinement, converging on a final product through a series of increasingly precise steps. See also *epigenesis.*

superior colliculus. Layered structure at the roof of the midbrain; receives visual input from the retina. See also *midbrain, retina.*

synapse. A connection between neurons in which one cell almost touches another, allowing neurotransmitters to pass between them. See also *neurotransmission, neurotransmitter.*

synaptic strengthening. The process by which connections between neurons are strengthened. See also *synapse.*

tectum. A sensory way station in the midbrain. See also *midbrain.*

template, protein. See *protein template.*

thalamus. A neural way station between sensory systems and cortical brain areas that further analyzes sensory input; contains the lateral geniculate nucleus and the medial geniculate nucleus. See also *medial geniculate nucleus.*

topographic map. A systematically ordered set of connections from one brain area to another, such as from the retina to the tectum or visual thalamus. See also *retina, tectum, thalamus.*

transcribed. See *transcription, gene.*

transcription, gene. The first stage in gene expression, a process by which DNA is copied onto an RNA complement. See also *DNA, gene expression, RNA.*

transcription factor. See *regulatory protein.*

transgenic animal. An animal whose genome has been genetically engineered, such as a knockout mouse. See also *knockout mouse.*

translation. The process that constructs sequences of amino acids based on the information contained in sequences of RNA codons. See also *amino acid.*

tree structure. A hierarchical representation of a sentence (linguistics).

twin, identical. Twins grown from a single fertilized egg, often, though not always, in a single placenta. See also *twins, fraternal.*

twins, fraternal. Twins grown from two separate fertilized eggs in separate placentas. See also *twins, identical.*

Universal Grammar (UG). Aspects of language that are shared across all languages and all speakers; Noam Chomsky's term for the innate contribution to language. See also *language acquisition device.*

vasopressin. A hormone released by the pituitary gland that acts on the kidney and plays a role in social behavior.

vertebrate. Any animal with a backbone, including fish, amphibians, birds, reptiles, and mammals. See also *invertebrate.*

visual cortex. Part of the occipital lobe devoted to visual processing.

vocal tract. Organs of speech, such as the lips, tongue, vocal folds, and trachea.

Williams syndrome. A form of mental retardation with a complex pattern of intact and impaired abilities; linked to a deletion of a small region of Chromosome 7.

X chromosome. One of the two sex chromosomes; females have two X chromosomes; males have an X and a Y. See *Y chromosome.*

Y chromosome. The "male" sex chromosome. See *X chromosome.*

zygote. A fertilized egg.

NOTES

CHAPTER 1   NEITHER IS BETTER

1. Crick, 1993.
2. Pinker, 1997.
3. Zatorre, 2001.
4. Phan, Wager, Taylor, & Liberzon, 2002.
5. Tiihonen et al., 1994.
6. Ehrlich, 2000.
7. Ibid.
8. Menand, 2002.
9. Watson, 1925.
10. DeFries, Gervais, & Thomas, 1978.
11. http://abcnews.go.com/sections/world/DailyNews/finland020306.html.
12. Thompson et al., 2001.
13. Posthuma et al., 2002.
14. Bartley, Jones, & Weinberger, 1997; Lohmann, von Cramon, & Steinmetz, 1999.
15. Corpus callosum: Scamvougeras, Kigar, Jones, Weinberger, & Witelson, 2003; brain structures more generally: Pennington et al., 2000.
16. Kaschube, Wolf, Geisel, & Lowel, 2002.
17. Meltzoff & Moore, 1977.
18. Morrongiello, Fenwick, & Chance, 1998.
19. Ramus, Hauser, Miller, Morris, & Mehler, 2000.
20. Farroni, Csibra, Simion, & Johnson, 2002.
21. Pinker, 1994.
22. Dehaene, 1997.
23. Pinker, 2002.
24. Nelkin, 2001.
25. Bateson, 2001.
26. Hogenesch et al., 2001; International Human Genome Sequencing Consortium, 2001; Venter et al., 2001.
27. Kandel, Schwartz, & Jessell, 2000.

28. Goodman, 1978; Loer, Steeves, & Goodman, 1983.

29. Biondi et al., 1998; Bonan et al., 1998; Thompson et al., 2001.

30. http://www.boyakasha.co.uk/women.mp3.

31. Plomin, 1997.

32. In the case of identical twins, the heritability of a given trait is easily calculated. Simply find the difference between the correlation between identical twins and the correlation between nonidentical twins and double it. For instance, if the correlation between the height of one identical twin is strongly correlated (0.8) with the height of the other, but the height of one fraternal is more moderately correlated (0.6) with the height of the other, the difference between correlations would be 0.2, which, when multiplied by 2, would yield a heritability of 0.4 (40 percent). Plomin, 1997.

33. Bouchard, 1994; Plomin, 1997.

34. Bouchard & McGue, 1981; Rowe, 1994.

35. Lykken, 1982.

36. Bouchard & Loehlin, 2001; Plomin, DeFries, McClearn, & McGuffin, 2001.

37. Block, 1996.

38. Ibid.

39. J. Halberda, personal communication, February 2003.

40. Heritability scores also inevitably reflect the range of genetic variability in the populations from which they are drawn. If gene X increases growth in conjunction with gene A_1 but decreases growth in conjunction with gene variant A_2, estimates of gene X's heritability will vary as a function of the distribution of A_1 versus A_2—not simply as a function of what X does, but as a function of how X interacts with A_1 and A_2.

41. Medawar, 1981.

42. Bateson, 2002.

43. Ten billion neurons: Blinkov & Glezer, 1968; 30,000 genes: International Human Genome Sequencing Consortium, 2001; Venter et al., 2001.

44. King & Wilson, 1975.

CHAPTER 2   BORN TO LEARN

1. *Rock Hill Herald*: January 1, 1999.

2. http://www.theonion.com/onion3119/stupidbabies.html.

3. Chomsky, 2000; Chomsky, 1965.

4. Piaget, 1954.

5. Ibid.

6. Fantz, 1961.

7. Baillargeon, Spelke, & Wasserman, 1985.

8. Bogartz, Shinskey, & Speaker, 1997; Rivera, Wakeley, & Langer, 1999.

9. Wynn, 1992; Cohen & Marks, 2002; Wynn, 2002.

10. Munakata, McClelland, Johnson, & Siegler, 1997.

11. Marcus & Clifton, in preparation; Smith, Thelen, Titzer, & McLin, 1999.

12. Martin, 1998.

13. Clifton, Perris, & McCall, 1999; Clifton, Rochat, Litovsky, & Perris, 1991.

14. Goren, Sarty, & Wu, 1975; Johnson & Morton, 1991.

15. Farroni et al., 2002.

16. Nazzi, Floccia, & Bertoncini, 1998; Nazzi, Bertoncini, & Mehler, 1998; Ramus et al., 2000.

17. Vouloumanos & Werker, submitted.

18. Not all improvement is the result of experience, By the time they are six months old, infants can tell familiar faces from unfamiliar ones. But that's true not just for human faces, but also for the ability of human infants to recognize *monkey* faces, an amazing ability, perhaps a leftover from an earlier era, that vanishes as children get older (Pascalis, de Haan, & Nelson, 2002).

19. Regolin, Vallortigara, & Zanforlin, 1995.

20. Regolin, Tommasi, & Vallortigara, 2000.

21. Coppinger & Coppinger, 2001.

22. Hall, 1994.

23. Baker, Taylor, & Hall, 2001.

24. Sachs, 1988.

25. Fentress, 1973.

26. Greer & Capecchi, 2002.

27. Larsen, Vestergaard, & Hogan, 2000.

28. Hauser, 2002.

29. Marler, 1991.

30. Kandel, 1979; Rankin, 2002.

31. Watson is widely known for an experiment that allegedly established the power of such simple learning techniques in humans. According to legend, Watson loudly struck "a hammer upon a [four-foot-long] suspended steel bar" right behind the head of an eleven-month-old child named Albert every time Albert approached the rat, thereby inducing in him a terrible fear of white rats. In recent years, however, the soundness and veridicality of this study has come into considerable question (Harris, 1979; Paul & Blumenthal, 1989).

32. Gallistel, 1990; Gallistel, Brown, Carey, Gelman, & Keil, 1991.

33. Emlen, 1975.

34. About a decade before the term "virtual reality" was invented, Cornell ecologist Stephen Emlen raised a set of buntings in a specially designed planetarium in which everything rotated not around Polaris (the current center of the heavens) but around Betelgeuse (one of the brighter stars of the Southern Hemisphere). The poor birds oriented themselves backward precisely as if they mistook Betelgeuse for North—proof positive that buntings rely on the rotation of the heavens rather than a built-in compass.

35. Gallistel, 1990; Renner, 1960.

36. Marler, 1984.

37. http://www.stanfordalumni.org/birdsite/text/species/Willow_Flycatcher.html.

38. Kroodsma, 1984.

39. http://www.bird-friends.com/NorthernMockingbird.html.

40. Smith, King, & West, 2000.

41. Saffran, Aslin, & Newport, 1996.

42. Marcus, 2000; Marcus, Vijayan, Bandi Rao, & Vishton, 1999, and similar experiments by Gomez & Gerken, 1999.

43. Hauser, Weiss, & Marcus, 2002.

44. Rizzolatti, Fadiga, Gallese, & Fogassi, 1996.

45. Tomasello, 1999; Caldwell & Whiten, 2002.

46. Meltzoff & Moore, 1977.

47. Whiten et al., 1999.

48. Van Schaik et al., 2003.

49. Richerson & Boyd, 2004.

50. Hare, Brown, Williamson, & Tomasello, 2002; Povinelli, 2000.

51. Bloom, 2000.

52. Cheney & Seyfarth, 1990; Seyfarth, Cheney, & Marler, 1980.

53. Savage-Rumbaugh et al., 1993.

54. Fenson et al., 1994.

55. Bloom, 2000; Dromi, 1987.

56. Carey, 1978; Pinker, 1994.

57. Liittschwager & Markman, 1994; Markman, 1989.

58. Brown, 1957.

59. Pinker, 1994.

60. Savage-Rumbaugh et al., 1993.

61. Heath, 1983; Ochs & Schieffelin, 1984.

62. Huttenlocher, 1998.

63. Chomsky, 1975; Pinker, 1994.

CHAPTER 3   BRAIN STORMS

1. DeCasper & Spence, 1986.

2. Moon & Fifer, 2000.

3. Ibid.

4. Hubel & Wiesel, 1962; Wiesel & Hubel, 1963.

5. Hubel, 1988. The mechanism of this particular bit of tuning is competition: If one eye is deprived, the other takes over some of the real estate that was devoted to the deprived eye, ultimately withering the neat columns of ocular dominance. If both eyes are deprived, competition favors neither eye and the ocular dominance columns survive.

6. Gödecke & Bonhoeffer, 1996.

7. Law & Constantine-Paton, 1981.

8. Dunlop, Lund, & Beazley, 1996.

9. Internally generated activity and three-eyed frogs: Reh & Constantine-Paton, 1985.

10. Crowley & Katz, 1999.

11. Miyashita-Lin, Hevner, Wassarman, Martinez, & Rubenstein, 1999.

12. Verhage et al., 2000.

13. Molnar et al., 2002; Washbourne et al., 2002.

14. O'Leary & Stanfield, 1989; Stanfield & O'Leary, 1985.

15. Deacon, 2000; Isacson & Deacon, 1997.

16. Webster, Ungerleider, & Bachevalier, 1995.

17. Sur & Leamey, 2001; Sur, Pallas, & Roe, 1990.

18. Rebillard, Carlier, Rebillard, & Pujol, 1977.

19. Rauschecker, 1995.

20. Bavelier & Neville, 2002; Neville & Lawson, 1987.

21. Alho, Kujala, Paavilainen, Summala, & Naatanen, 1993; Kujala, Alho, & Naatanen, 2000.

22. Sadato et al., 1996.

23. Vargha-Khadem, Isaacs, & Muter, 1994.

24. De Bode & Curtiss, 2000; Vicari et al., 2000.

25. Merzenich et al., 1984.

26. Bates, 1999.

27. Quartz & Sejnowski, 1997.

28. Gould, Reeves, Graziano, & Gross, 1999; Rakic, 1998.

29. Goldman-Rakic, Bourgeois, & Rakic, 1997; Huttenlocher, 1990.

30. Thulborn, Carpenter, & Just, 1999.

31. Saunders, 1982.

32. Bjornson, Rietze, Reynolds, Magli, & Vescovi, 1999.

33. Angelucci, Clasca, & Sur, 1998.

34. Ibid.

35. Bates, 1999.

36. Balaban, 1997.

37. Levitt, 2000.

38. Johnson, 1997; Pinker, 2002.

39. Ledoux, 1996.

40. Damasio, 1994.

41. Sur & Leamey, 2001.

42. Kandel, Schwartz, & Jessell, 2000.

43. Pinker, 2002.

44. Kaas, 2002; Pinker, 2002.

45. Scoville & Milner, 1957.

46. Vargha-Khadem et al., 1997.

47. Maruishi et al., 2001.

48. Ballaban-Gil, Rapin, Tuchman, & Shinnar, 1996.

49. Shaywitz et al., 1999.

CHAPTER 4 ARISTOTLE'S IMPETUS

1. Successive approximation in the development of an embryo: O'Rahilly, Müller, & Streeter, 1987; http://embryo.soad.umich.edu/resources/morph.mov; http://www.visembryo.com/.

2. Gould, 1977.

3. Spock, 1957.

4. Gould, 1977; Richardson et al., 1998; http://zygote.swarthmore.edu/evo5.html.

5. Shell, 2002.

6. http://www.sonic.net/~nbs/projects/anthro201/disc/.

7. Mendel's factors might be more properly thought of as *alleles,* variants of particular genes. For example, when scientists refer to a gene for cystic fibrosis, what they mean is that a particular version (allele) of a gene that *all* people have is corrupted in that particular individual.

8. Oetting & Bennett, 2003.

9. Purves, Sadava, Orians, & Heller, 2001.

10. Gasking, 1959.

11. Garrod, 1923.

12. Beadle & Tatum, 1941; Morange, 1998.

13. Although it is common to think of each gene as corresponding to only a single protein, because of processes such as "alternative splicing," which I will describe in Chapter 8, there are actually perhaps ten times as many proteins in the human body than there are genes in our genome.

14. Alberts et al., 1994.

15. Tanford & Reynolds, 2001; Walsh, 2002.

16. Judson, 1979.

17. Olby, 1994.

18. Chargaff, 1950.

19. Ibid.

20. Hershey & Chase, 1952.

21. Judson, 1979.

22. http://www.geocities.com/jenaith/DNA1.html.

23. Klug, 1974; Sayre, 1975.

24. Watson & Crick, 1953; Judson, 1979.

25. Watson & Crick, 1953.

26. Judson, 1979.

27. Nesse & Williams, 1994; Rensberger, 1996.

28. *E. coli* has a bad reputation, but most strains are beneficial. Stomach distress comes not from too much *E. coli* but from too much of a particular strain of *E. coli: 0157:H7,* bane of the food industry and an excellent reason to wash your hands after handling meat.

29. Jacob & Monod, 1961; Judson, 1979.

30. Wilmut, Schnieke, McWhir, Kind, & Campbell, 1997.

31. Hsiao et al., 2001.

32. Alberts et al., 1994; Davidson, 2001; Furlow & Brown, 1999; Marks, Iyer, Cui, & Merchant, 1996.

33. Beldade & Brakefield, 2002; Brakefield et al., 1996.

34. Sydney Brenner, as quoted in Gehring, 1998.

35. Gilbert, 2000; Kimble & Austin, 1989.

36. Saunders, Gasseling, & Cairns, 1959.

CHAPTER 5   COPERNICUS'S REVENGE

1. Restak, 1979.
2. Rose, 1973.
3. Davis, 1997.
4. Restak, 1979.
5. Montgomery, 1999; Seeman & Madras, 2002.
6. Kandel, Schwartz, & Jessell, 2000.
7. McIlwain & Bachelard, 1985.
8. Koch & Segev, 2000.
9. Feldman & Ballard, 1982; Kandel, Schwartz, & Jessell, 2000; Oram & Perrett, 1992.
10. Hall, 1992.
11. Kandel, Schwartz, & Jessell, 2000.
12. Alberts et al., 1994.
13. Lequin & Barkovich, 1999; Ross & Walsh, 2001.
14. Masland, 2001.
15. Anderson, 1992.
16. Nobody is exactly sure why nature follows this strategy rather than just producing the "right" number of cells in the first place. The generate-and-cull strategy could simply be a holdover of evolution—as engineers realize, it is often easier to tweak an existing system than to start from scratch—but it might also be a way of building in developmental flexibility. In complex organisms where there are millions or billions of neurons, it would foolish to count on any particular neuron to get a particular job done. By separating out the process of neuron production from the process by which neurons ultimately become committed to their fate, nature may achieve an extra level of flexibility and robustness.
17. Ikonomidou et al., 2001.
18. Rakic, 1972.
19. Marin & Rubenstein, 2003.
20. Harris, Honigberg, Robinson, & Kenyon, 1996.
21. Maynard Smith & Szathmáry, 1995.
22. Chenn & Walsh, 2002.
23. Stuhmer, Anderson, Ekker, & Rubenstein, 2002.
24. Marin, Yaron, Bagri, Tessier-Lavigne, & Rubenstein, 2001.
25. Fukuchi-Shimogori & Grove, 2001.
26. Crossley, Martinez, Ohkubo, & Rubenstein, 2001.
27. Bishop, Goudreau, & O'Leary, 2000; Mallamaci, Muzio, Chan, Parnavelas, & Boncinelli, 2000; Tole, Goudreau, Assimacopoulos, & Grove, 2000.
28. Ross & Walsh, 2001; Webb, Parsons, & Horwitz, 2002.
29. Hsiao et al., 2001; Warrington, Nair, Mahadevappa, & Tsyganskaya, 2000.
30. Mattson, 2002; Roth & D'sa, 2001.
31. International Human Genome Sequencing Consortium, 2001; Patthy, 2003; Venter et al., 2001.
32. Hsiao et al., 2001; Warrington, Nair, Mahadevappa, & Tsyganskaya, 2000.

33. For a discussion of some of the techniques used in genetics and developmental neuroscience, see the Appendix.

34. Sokolowski, 1998.

35. Young, Nilsen, Waymire, MacGregor, & Insel, 1999.

36. Murphy et al., 2001.

37. Gross et al., 2002.

38. Greer & Capecchi, 2002.

39. Lefebvre et al., 1998.

40. Gainetdinov et al., 1999.

41. Grimsby et al., 1997.

42. Sillaber et al., 2002.

43. http://www.ncbi.nlm.nih.gov/omim/.

44. Fisher & DeFries, 2002; Plomin & McGuffin, 2003.

45. Egan et al., 2003.

46. Caspi et al., 2002.

47. Hariri et al., 2002.

48. Tecott, 2003; Waterston et al., 2002.

49. Waterston et al., 2002.

50. International Human Genome Sequencing Consortium, 2001; Venter et al., 2001.

51. Gehring, 2002.

52. Kooy, 2003.

53. Even though human genes often have mouse counterparts (and vice versa), a given gene may well have a different effect in a person than it does in a mouse. Serotonin, for example, may modulate anxiety in both mice and people, but humans might have additional systems for modulating anxiety that are not available in a mouse brain. Animal models can give valuable hints as to the specific functions of particular genes in humans but provide no guarantee that the operation of those genes will be identical.

54. Pennartz, Uylings, Barnes, & McNaughton, 2002.

55. Bernheim & Mayeri, 1995; Scheller & Axel, 1984.

56. Anand et al., 2001; Baker, Taylor, & Hall, 2001; Emmons & Lipton, 2003.

57. Anand et al., 2001; Baker, Taylor, & Hall, 2001; Song et al., 2002.

58. Lykken, McGue, Tellegen, & Bouchard, 1992.

59. Karmiloff-Smith, 1998.

60. Liu, Dwyer, & O'Leary, 2000.

61. Fukuchi-Shimogori & Grove, 2001; O'Leary & Nakagawa, 2002.

62. Nakagawa & O'Leary, 2001; Sestan, Rakic, & Donoghue, 2001; Skeath & Thor, 2003.

63. Bessa, Gebelein, Pichaud, Casares, & Mann, 2002; Flores et al., 2000.

64. St. Francis quote from Saint Bonaventure's *Life of Francis*.

CHAPTER 6   WIRING THE MIND

1. Buchsbaum et al., 1998; Kumar & Cook, 2002.

2. Gazzaniga, 1998; Sperry, 1961.

3. http://www.indiana.edu/~pietsch/split-brain.html.

4. Kullander et al., 2003.

5. Emmons & Lipton, 2003.

6. White, Southgate, Thomson, & Brenner, 1986.

7. Sanes, Reh, & Harris, 2000.

8. Hibbard, 1965.

9. Harris, 1986.

10. Harris, Holt, & Bonhoeffer, 1987.

11. Goodhill, 1998.

12. Clandinin & Zipursky, 2002; Redies, 2000.

13. Rajagopalan, Vivancos, Nicolas, & Dickson, 2000; Simpson, Bland, Fetter, & Goodman, 2000.

14. The set of growth cones on a given receptor is by no means fixed. For example, in many fly axons (and likely also in many human axons), the number of *Robo* receptors starts low and then radically increases after the axons cross a central left-right divide, a neat trick that allows a growth cone to break its journey up into more easily navigated chunks.

15. Sharma, Leonard, Lettieri, & Pfaff, 2000.

16. Komiyama, Johnson, Luo, & Jefferis, 2003.

17. Kaas, 2002; Welker, 2000.

18. Welker, 2000.

19. Kaas, 2002; Welker, 2000.

20. Huffman et al., 1999.

21. Sherry, Jacobs, & Gaulin, 1992.

22. Finlay, Darlington, & Nicastro, 2001.

23. Davis & Squire, 1984; Fields, Eshete, Stevens, & Itoh, 1997; Flexner, Flexner, & Stellar, 1965; Kaczmarek, 1993; Kandel & O'Dell, 1992; Wallace et al., 1995.

24. Castren, Zafra, Thoenen, & Lindholm, 1992; Kaczmarek, Zangenehpour, & Chaudhuri, 1999; Rosen, McCormack, Villa-Komaroff, & Mower, 1992.

25. Hofmann, 2003.

26. Fields, Eshete, Dudek, Ozsarac, & Stevens, 2001; Kaczmarek, 2000.

27. Behar et al., 2001; Hannan et al., 2001; Lopez-Bendito et al., 2002; Metin, Denizot, & Ropert, 2000; Zhang & Poo, 2001.

28. Al-Majed, Brushart, & Gordon, 2000; Morimoto, Miyoshi, Fujikado, Tano, & Fukuda, 2002; Song, Zhao, Forrester, & McCaig, 2002.

29. Hebb, 1947; Klintsova & Greenough, 1999.

30. Rampon et al., 2000.

31. Gustafsson & Kraus, 2001.

32. Black, Isaacs, Anderson, Alcantara, & Greenough, 1990.

33. Dubnau, Chiang, & Tully, 2003; Martin, Grimwood, & Morris, 2000.

34. Sanes & Lichtman, 1999.

35. Rose, 2000.

36. Dubnau, Chiang, & Tully, 2003; Kandel, 2001.

37. Tsien, Huerta, & Tonegawa, 1996.

38. Silva, Paylor, Wehner, & Tonegawa, 1992; Silva, Stevens, Tonegawa, & Wang, 1992.

39. Rose, 2000.

40. Chew, Mello, Nottebohm, Jarvis, & Vicario, 1995.

41. Tang et al., 1999.

42. Dubnau, Chiang, Grady et al., 2003.

43. Wei et al., 2001.

44. Dubnau, Chiang, & Tully, 2003.

45. Gallistel, 2002.

46. Ibid.; Martin, Grimwood, & Morris, 2000; Pena De Ortiz & Arshavsky, 2001; Shors & Matzel, 1997.

47. Dietrich & Been, 2001; Holliday, 1999; Pena De Ortiz & Arshavsky, 2001.

48. Buckner, Kelley, & Petersen, 1999; Fuster, 2000.

49. Ledoux, 1996.

50. Schacter, 1996.

51. Ramos, 2000.

52. Ledoux, 1996.

53. Mayford et al., 1996.

54. Bolhuis & Honey, 1998; Horn & McCabe, 1984; Johnson, Bolhuis, & Horn, 1985.

55. Bottjer, Miesner, & Arnold, 1984.

56. Nottebohm, Stokes, & Leonard, 1976; Wild, 1997.

57. Mello, Vicario, & Clayton, 1992.

58. Morrison & van der Kooy, 2001.

59. Hobert, 2003; Rankin, 2002.

60. Morrison, Wen, Runciman, & van der Kooy, 1999.

61. Hobert, 2003.

62. Costa et al., 2002.

63. Coppinger & Coppinger, 2001.

64. Johnson & Newport, 1989; Lenneberg, 1967.

65. Gregersen, Kowalsky, Kohn, & Marvin, 2001; Schlaug, 2001.

66. Knudsen & Knudsen, 1990; Linkernhoker & Knudsen, 2002.

67. Merzenich et al., 1984.

68. Daw, 1994.

69. Pizzorusso et al., 2002.

70. Bradbury et al., 2002.

71. Penn & Shatz, 1999.

72. Galli & Maffei, 1988.

73. Katz & Shatz, 1996; Penn & Shatz, 1999; Stellwagen & Shatz, 2002; Wong, 1999.

74. Katz & Shatz, 1996; Mastronarde, 1983.

75. Meister, Wong, Baylor, & Shatz, 1991.

76. Penn & Shatz, 1999.

77. Garaschuk, Linn, Eilers, & Konnerth, 2000; Momose-Sato, Miyakawa, Mochida, Sasaki, & Sato, 2003; Nedivi, Hevroni, Naot, Israeli, & Citri, 1993.

78. Schmidt & Eisele, 1985.

79. Weliky & Katz, 1997.
80. Stellwagen & Shatz, 2002.

CHAPTER 7   THE EVOLUTION OF MENTAL GENES

1. Seidl, Cairns, & Lubec, 2001.
2. Bowmaker, 1998.
3. Jameson, Highnote, & Wasserman, 2001.
4. De Duve, 1995; Holland, 1997.
5. Lacalli, 2001; Shu et al., 1992.
6. Lawn, Mackie, & Silver, 1981; Leys, Mackie, & Meech, 1999.
7. Ion channels themselves may actually have derived from a class of "transporter" proteins that controlled the flow of ions, one at a time, by enveloping each one individually and then carrying it across the membrane (Harris-Warrick, 2000).
8. Jegla & Salkoff, 1994.
9. Harris-Warwick, 2000.
10. Ortells & Lunt, 1995; Wo & Oswald, 1995; Xue, 1998.
11. Harris-Warrick, 2000; Wessler, Kirkpatrick, & Racke, 1999.
12. Fans of the late Stephen Jay Gould will realize that I am being a bit loose and anthropocentric here. Evolution has no particular goal, and bacteria have been evolving all the while, just as our ancestors have. Signaling systems improved over the course of our ancestry but in other lines may have gotten worse. We make our living through vastly complex systems for internal signaling, but that doesn't mean all other creatures do. When I say something has improved or evolved, I speak only of our own particular path on the tree of life.
13. Allman, 1999.
14. Jegla & Salkoff, 1994.
15. Ruiz-Trillo, Riutort, Littlewood, Herniou, & Baguna, 1999.
16. Sarnat & Netsky, 1985.
17. Younossi-Hartenstein, Jones, & Hartenstein, 2001.
18. Bullock, Moore, & Fields, 1984.
19. Richardson, Pringle, Yu, & Hall, 1997.
20. Wells, 1966.
21. Allman, 1999.
22. Oksenberg, Barcellos, & Hauser, 1999.
23. Carroll, Grenier, & Weatherbee, 2001; Gehring, 1998; Gerhart & Kirschner, 1997.
24. Garcia-Fernandez & Holland, 1994; Lacalli, 2001.
25. Allman, 1999; Manzanares et al., 2000.
26. Schad, 1993.
27. Butler & Hodos, 1996.
28. De Winter & Oxnard, 2001.
29. Simeone, Puelles, & Acampora, 2002; Williams & Holland, 2000.
30. Simeone, Puelles, & Acampora, 2002.

31. Mombaerts, 1999.

32. Costagli, Kapsimali, Wilson, & Mione, 2002; Patthy, 2003.

33. Rice & Curran, 2001.

34. Fatemi, 2002.

35. Jaaro, Beck, Conticello, & Fainzilber, 2001.

36. Frank & Kemler, 2002; Hilschmann et al., 2001; Pena De Ortiz & Arshavsky, 2001.

37. Patthy, 2003; Venter et al., 2001.

38. Kaas, 1987.

39. Krubitzer & Huffman, 2000.

40. Kaas, 1987.

41. Tavare, Marshall, Will, Soligo, & Martin, 2002.

42. Krubitzer, 2000.

43. Clancy, Darlington, & Finlay, 2001.

44. Carruthers & Boucher, 1998; Gumperz & Levinson, 1996; Levinson, Kita, Haun, & Rasch, 2002; Whorf, 1975[1956].

45. Darwin, 1874.

46. Fodor, 1975; Pinker, 1994.

47. Pinker, 1994.

48. Li & Gleitman, 2002.

49. Gleitman, Gleitman, Miller, & Ostrin, 1996; Tversky & Gati, 1982.

50. Ellis & Hennelly, 1980; Hoosain & Salili, 1987.

51. Davidoff, Davies, & Roberson, 1999; Roberson, Davies, & Davidoff, 2000.

52. Many studies of the relationship between language and thought have assumed that if some parts of thought are conducted in language, speakers of different languages should necessarily conceptualize the world in fundamentally different ways, but if (as has often been suggested on independent grounds) all languages express much the same range of thoughts, we might not expect there to be profound cognitive differences between speakers of different languages, after all.

53. Morgan, 1995; Verhaegen, 1988.

54. Corballis, 1992.

55. Lieberman, 1984.

56. Gould, 1979.

57. Pinker & Bloom, 1990.

58. Dunbar, 1996.

59. Miller, 2000.

60. Goodglass, 1993.

61. Embick, Marantz, Miyashita, O'Neil, & Sakai, 2000; Goodglass, 1993; Stromswold, Caplan, Alpert, & Rauch, 1996.

62. Goodglass, 1993; Wise et al., 2001.

63. Koelsch et al., 2002.

64. Friederici, 2002; Kaan & Swaab, 2002.

65. Crosson, 1992; Lieberman, 2002; Nadeau & Crosson, 1997.

66. Damasio, Grabowski, Tranel, Hichwa, & Damasio, 1996; Martin, Haxby, Lalonde, Wiggs, & Ungerleider, 1995; Martin, Wiggs, Ungerleider, & Haxby, 1996.

67. Pulvermuller, 2002.

68. Dronkers, 2000.

69. Rilling & Insel, 1999; Semendeferi & Damasio, 2000.

70. Jerison, 1979.

71. Coppinger & Coppinger, 2001; Coren, 1994.

72. Stephan, Frahm, & Baron, 1981.

73. Wickett, Vernon, & Lee, 2000.

74. Hedges & Nowell, 1995.

75. Kranzler, Rosenbloom, Martinez, & Guevara-Aguirre, 1998; Lenneberg, 1967; Skoyles, 1999.

76. Finlay, Darlington, & Nicastro, 2001.

77. Coppinger & Coppinger, 2001.

78. Cantalupo & Hopkins, 2001; Gannon, Holloway, Broadfield, & Braun, 1998.

79. Rilling & Insel, 1999.

80. Ibid.

81. Jeeves & Temple, 1987.

82. Geschwind & Levitsky, 1968; Ojemann, 1993; Shapleske, Rossell, Woodruff, & David, 1999.

83. Buxhoeveden, Switala, Litaker, Roy, & Casanova, 2001.

84. Strictly speaking, one could have a mind that was innate without having modules, but in the technical literature the two hypotheses have often been treated as interdependent, in part because the best current hypothesis about what it would be for the mind to have innate structure is for it to have innate *modular* structure.

85. Logan & Tabin, 1999; Margulies, Kardia, & Innis, 2001; Takeuchi et al., 1999.

86. Hauser, Weiss, & Marcus, 2002.

87. Postle & Corkin, 1998.

88. Gilbert, 2000.

89. Fisher & DeFries, 2002; Ramus et al., 2003.

90. Stein, 2001.

91. Temple et al., 2003.

92. van der Lely, Rosen, & McClelland, 1998.

93. van der Lely & Stollwerck, 1996.

94. A complication here is that even if there really are disorders that selectively impair whatever circuitry is special to language, there may be nothing to prevent people so impaired from using general cognitive capacities to fill in for missing circuitry; such "compensatory" strategies are one of the chief challenges in the study of the psychology of humans.

95. Hare, Brown, Williamson, & Tomasello, 2002.

96. Bloom, 2001.

97. Baldwin, 1991.

98. Baron-Cohen, Leslie, & Frith, 1985; Wimmer & Perner, 1983.

99. Onishi & Baillargeon, 2002.

100. Call & Tomasello, 1999.

101. Tomasello, Call, & Hare, 2003.

102. Marcus, 2001a.

103. Hauser, Chomsky, & Fitch, 2002.

104. Fauconnier & Turner, 2002.

105. Lieberman, 2002.

106. Fitch, 2000.

107. Bickerton, 1990; Deacon, 1997; Jackendoff, 2002; Pinker & Bloom, 1990.

108. Jackendoff, 2002.

109. Chen & Li, 2001; Brunet et al., 2002.

110. Tavare, Marshall, Will, Soligo, & Martin, 2002.

111. Mojzsis et al., 1996.

112. Wilson & Sarich, 1969.

113. Diamond, 1992.

114. Hall et al., 1990.

115. Sagan & Druyan, 1992.

116. Ebersberger, Metzler, Schwarz, & Paabo, 2002.

117. Ibid.

118. Ioshikhes & Zhang, 2000.

119. Fauconnier & Turner, 2002; Klein & Edgar, 2002; Mithen, 1996.

120. Fisher, Vargha-Khadem, Watkins, Monaco, & Pembrey, 1998; Lai, Fisher, Hurst, Vargha-Khadem, & Monaco, 2001.

121. Gopnik & Crago, 1991; Vargha-Khadem, Watkins, Alcock, Fletcher, & Passingham, 1995; Watkins, Dronkers, & Vargha-Khadem, 2002.

122. Vargha-Khadem, Watkins, Alcock, Fletcher, & Passingham, 1995.

123. Lai, Fisher, Hurst, Vargha-Khadem, & Monaco, 2001.

124. Shu, Yang, Zhang, Lu, & Morrisey, 2001.

125. Enard et al., 2002; Zhang, Webb, & Podlaha, 2002.

126. Enard et al., 2002; Zhang, Webb, & Podlaha, 2002.

127. Boyd & Silk, 2000; Klein & Edgar, 2002.

CHAPTER 8 PARADOX LOST

1. Aubin, Dery, Lemieux, Chailler, & Jeannotte, 2002.

2. This sort of flexibility is not without cost—some cancers, for example, are the result of mature cells incorrectly taking up their developmental programs a second time.

3. Brockes, 1997.

4. Mey, 2001.

5. Cebria, Kobayashi et al., 2002; Cebria, Nakazawa et al., 2002.

6. Dunlop, Tran, Tee, Papadimitriou, & Beazley, 2000.

7. Dunlop et al., 2002.

8. Watkins & Barres, 2002.

9. From the perspective of evolution, the benefit of neural regeneration in complex mammals might have been relatively small, since until recently, an organism with a serious head injury had little chance of survival.

10. Profet, 1992.

11. Cruz, 1997.

12. http://www.tidbits.com/netbits/nb-issues/NetBITS–003.html.

13. http://www.amara.com/IEEEwave/IW_fbi.html.

14. Pentland, 1997.

15. http://www.nhgri.nih.gov/educationkit/basicGenetics.html.

16. For the latest estimate of the number of human genes, point your web browser to http://www.ensembl.org/Homo_sapiens/stats/; for a touch of humility, follow that up with a visit to http://www.ensembl.org/Mus_musculus/stats/.

17. Hogenesch et al., 2001; International Human Genome Sequencing Consortium, 2001; Venter et al., 2001; and personal communications, May 2003, with the Ensembl Helpdesk and Michael Cooke.

18. Modrek, Resch, Grasso, & Lee, 2001.

19. Schmucker et al., 2000.

20. Brown et al., 2000.

21. Goodhill & Richards, 1999.

22. Marcus, 2001b.

23. Computer simulations of neurons are to real neurons what SimCity people are to real people: vast oversimplifications, but still perhaps enough to allow scientists some insight into how real neurons work.

24. Marcus, 2001b.

CHAPTER 9   FINAL FRONTIERS

1. Godwin, Luckenbach, & Borski, 2003.

2. Bishop, 2001; Johnson & Carey, 1998; Korenberg et al., 2000; van der Lely, Rosen, & McClelland, 1998.

3. Already, there are preliminary hints that the genes for learning may overlap with the genes for brain growth, much as the genes for regenerating lost limbs in a salamander overlap the genes for the salamander's initial limb growth. For example, early response genes may trigger the same synapse-formation genes that the brain uses in the development of other synapses that form prior to experience. Such overlap wouldn't mean that learning and brain growth would then be indistinguishable (any more than growth and regeneration are indistinguishable in the salamander), but it would illustrate one more way in which nature gets the most out of the machinery it has, repurposing useful tools for many different situations.

4. Rossi & Cattaneo, 2002.

5. Hunt & Vorburger, 2002; Sapolsky, 2003.

6. Sapolsky, 2003.

7. Hutchinson, 2001.

8. Evans & Relling, 1999; Pagliarulo, Datar, & Cote, 2002; Tsai & Hoyme, 2002; http://www.nigms.nih.gov/pharmacogenetics/.

9. Mancama & Kerwin, 2003; Cacabelos, 2002; Basile, Masellis, Potkin, & Kennedy, 2002.

10. Winsberg & Comings, 1999.

11. Goho, 2003.

12. At present, bioengineers have the wherewithal to clone mammals, but there are still some significant bugs in the system, perhaps having to do with aging-related telomeres at the ends of chromosomes and/or with maternally derived "epigenetic" information about which genes should be switched on or off early in development. For further discussion, see Hochedlinger & Jaenisch, 2002.

13. Fukuyama (2002) also suggested that our ethical values depend on a particular notion of what it is to be a human being, and that fiddling with human nature could lead to ethical chaos; I refer you to his book for further discussion.

14. Gray, 2002.

15. http://www.kenanmalik.com/essays/pinker_gray.html.

16. http://www.sptimes.com/News/121799/Sports/Bannister_stuns_world.shtml.

17. Dickens & Flynn, 2001; Flynn, 1999.

18. Moore & Simon, 2000.

19. Caspi et al., 2002.

APPENDIX

1. Encarta, 1999 ed.

2. International Human Genome Sequencing Consortium, 2001; Venter et al., 2001.

3. Lebon, 2001.

4. Plomin & Crabbe, 2000.

5. Nasar, 1998.

6. Sherrington et al., 1988.

7. You might wonder why the dominant traits should be balanced 1:1 with the recessive traits, rather than outnumbering them 3:1 as in Mendel's experiment with the yellow and green peas. The difference is that Mendel was breeding recessives with pure-bred dominants, whereas Morgan was breeding his recessives with heterozygotes (that is, flies that inherited one copy of the dominant allele and one copy of the recessive allele).

8. Sturtevant, 1913.

9. Lee, Ladd, Bourke, Pagliaro, & Tirnady, 1994.

10. Part of the problem here is statistical, similar to one that pollsters routinely face. What you want to know is exactly how well one trait (having some particular marker) predicts another (say, manic-depression), but all you really know is how well one predicts the other *in your sample*. It may be true that 72 percent of your particular sample plans to vote for Dewey, but even so, Truman might still win. Just as it is hard to be sure how similar the rest of the population will be to your polling group, it is hard to tell how well the statistics of your sample predict the population as a whole, and any differences between the proportion of a trait in your sample and in the population as a whole will affect the accuracy of your linkage results.

11. http://www.ncbi.nlm.nih.gov/omim/.

12. Lander & Schork, 1994.

13. Ohno, 1970.

14. Eddy, 2001; Hirotsune et al., 2003.

15. Makalowski, 2000.

16. Morange, 2001.

17. Crawley, 2000.

18. Golling et al., 2002.

19. Hopkin, 2001; Husi & Grant, 2001; O'Donovan, Apweiler, & Bairoch, 2001.

20. "Genes Were Easy" was the subtitle of an April 2–4, 2001, conference in McLean, Virginia.

21. http://www.ncbi.nlm.nih.gov/BLAST/.

22. Carugo & Pongor, 2002; http://www.ncbi.nlm.nih.gov/entrez/query.fcgi?db=Structure.

REFERENCES

Al-Majed, A. A., Brushart, T. M., & Gordon, T. (2000). Electrical stimulation accelerates and increases expression of BDNF and trkB mRNA in regenerating rat femoral motoneurons. Eur J Neurosci, 12(12), 4381–4390.

Alberts, B., Bray, D., Lewis, J., Raff, M., Roberts, K., & Watson, J. D. (1994). Molecular biology of the cell, 3d ed. New York: Garland.

Alho, K., Kujala, T., Paavilainen, P., Summala, H., & Naatanen, R. (1993). Auditory processing in visual brain areas of the early blind: evidence from event-related potentials. Electroencephalogr Clin Neurophysiol, 86(6), 418–427.

Allman, J. M. (1999). Evolving brains. New York: Scientific American Library; distributed by W. H. Freeman.

Anand, A., Villella, A., Ryner, L. C., Carlo, T., Goodwin, S. F., Song, H. J., et al. (2001). Molecular genetic dissection of the sex-specific and vital functions of the Drosophila melanogaster sex determination gene fruitless. Genetics, 158(4), 1569–1595.

Anderson, D. J. (1992). Molecular control of neural development. In Z. W. Hall (ed.), An introduction to molecular neurobiology, pp. 355–387. Sunderland, Mass.: Sinauer Associates.

Angelucci, A., Clasca, F., & Sur, M. (1998). Brainstem inputs to the ferret medial geniculate nucleus and the effect of early deafferentation on novel retinal projections to the auditory thalamus. J Comp Neurol, 400(3), 417–439.

Aubin, J., Dery, U., Lemieux, M., Chailler, P., & Jeannotte, L. (2002). Stomach regional specification requires Hoxa5-driven mesenchymal-epithelial signaling. Development, 129(17), 4075–4087.

Baillargeon, R., Spelke, E. S., & Wasserman, S. (1985). Object permanence in five-month-old infants. Cognition, 20(3), 191–208.

Baker, B. S., Taylor, B. J., & Hall, J. C. (2001). Are complex behaviors specified by dedicated regulatory genes? Reasoning from Drosophila. Cell, 105(1), 13–24.

Balaban, E. (1997). Changes in multiple brain regions underlie species differences in a complex congenital behavior. Proc Natl Acad Sci USA, 94, 2001–2006.

Baldwin, D. (1991). Infant's contribution to the achievement of joint reference. Child Dev, 62, 875–890.

Ballaban-Gil, K., Rapin, I., Tuchman, R., & Shinnar, S. (1996). Longitudinal examination of the behavioral, language, and social changes in a population of adolescents and young adults with autistic disorder. Pediatr Neurol, 15(3), 217–223.

Baron-Cohen, S., Leslie, A. M., & Frith, U. (1985). Does the autistic child have a "theory of mind"? Cognition, 21(1), 37–46.

Bartley, A. J., Jones, D. W., & Weinberger, D. R. (1997). Genetic variability of human brain size and cortical gyral patterns. Brain, 120 (Pt 2), 257–269.

Basile, V. S., Masellis, M., Potkin, S. G., & Kennedy, J. L. (2002). Pharmacogenomics in schizophrenia: the quest for individualized therapy. Hum Mol Genet, 11(20), 2517–2530.

Bates, E. (1999). Plasticity, localization, and language development. In S. H. Broman & J. M. Fletcher (eds.), The changing nervous system: neurobiological consequences of early brain disorders. New York: Oxford University Press.

Bateson, P. (2001). Where does our behaviour come from? J Biosci, 26(5), 561–570.

____. (2002). The corpse of a wearisome debate. Science, 297(5590), 2212–2213.

Bavelier, D., & Neville, H. J. (2002). Cross-modal plasticity: where and how? Nat Rev Neurosci, 3(6), 443–452.

Beadle, G. W., & Tatum, E. L. (1941). Genetic control of biochemical reactions in Neurospora. Proc Natl Acad Sci USA, 27, 499–506.

Behar, T. N., Smith, S. V., Kennedy, R. T., McKenzie, J. M., Maric, I., & Barker, J. L. (2001). GABA(B) receptors mediate motility signals for migrating embryonic cortical cells. Cereb Cortex, 11(8), 744–753.

Beldade, P., & Brakefield, P. M. (2002). The genetics and evo-devo of butterfly wing patterns. Nat Rev Genet, 3(6), 442–452.

Bernheim, S. M., & Mayeri, E. (1995). Complex behavior induced by egg-laying hormone in Aplysia. J Comp Physiol [A], 176(1), 131–136.

Bessa, J., Gebelein, B., Pichaud, F., Casares, F., & Mann, R. S. (2002). Combinatorial control of Drosophila eye development by eyeless, homothorax, and teashirt. Genes Dev, 16(18), 2415–2427.

Bickerton, D. (1990). Language & species. Chicago: University of Chicago Press.

Bijeljac-Babic, R., Bertoncini, J., & Mehler, J. (1993). How do four-day-old infants categorise multisyllabic utterances? Dev Psychol, 29, 711–721.

Biondi, A., Nogueira, H., Dormont, D., Duyme, M., Hasboun, D., Zouaoui, A., et al. (1998). Are the brains of monozygotic twins similar? A three-dimensional MR study. Am J Neuroradiol, 19(7), 1361–1367.

Bishop, D. V. (2001). Genetic and environmental risks for specific language impairment in children. Philos Trans R Soc Lond B Biol Sci, 356(1407), 369–380.

Bishop, K. M., Goudreau, G., & O'Leary, D. D. (2000). Regulation of area identity in the mammalian neocortex by Emx2 and Pax6. Science, 288(5464), 344–349.

Bjornson, C. R., Rietze, R. L., Reynolds, B. A., Magli, M. C., & Vescovi, A. L. (1999). Turning brain into blood: a hematopoietic fate adopted by adult neural stem cells in vivo. Science, 283(5401), 534–537.

Black, J. E., Isaacs, K. R., Anderson, B. J., Alcantara, A. A., & Greenough, W. T. (1990). Learning causes synaptogenesis, whereas motor activity causes angiogenesis, in cerebellar cortex of adult rats. Proc Natl Acad Sci USA, 87(14), 5568–5572.

Blinkov, S. M., & Glezer, I. I. (eds.). (1968). The human brain in figures and tables. New York: Basic Books.

Block, N. (1996). How heritability misleads about race. Boston Review,, 20(6), 30–35.

Bloom, P. (2000). How children learn the meanings of words. Cambridge: MIT Press.

___. (2001). Precis of How children learn the meanings of words. Behav Brain Sci, 24(6), 1095–1103; discussion 1104–1134.

Bogartz, R. S., Shinskey, J. L., & Speaker, C. J. (1997). Interpreting infant looking: the event set X event set design. Dev Psychol, 33, 408–412.

Bolhuis, J. J., & Honey, R. C. (1998). Imprinting, learning and development: from behaviour to brain and back. Trends Neurosci, 21(7), 306–311.

Bonan, I., Argenti, A. M., Duyme, M., Hasboun, D., Dorion, A., Marsault, C., et al. (1998). Magnetic resonance imaging of cerebral central sulci: a study of monozygotic twins. Acta Genet Med Gemellol, 47(2), 89–100.

Bottjer, S. W., Miesner, E. A., & Arnold, A. P. (1984). Forebrain lesions disrupt development but not maintenance of song in passerine birds. Science, 224(4651), 901–903.

Bouchard, T. J., Jr. (1994). Genes, environment, and personality. Science, 264(5166), 1700–1701.

Bouchard, T. J., Jr., & Loehlin, J. C. (2001). Genes, evolution, and personality. Behav Genet, 31(3), 243–273.

Bouchard, T. J., Jr., & McGue, M. (1981). Familial studies of intelligence: a review. Science, 212(4498), 1055–1059.

Bowmaker, J. K. (1998). Evolution of colour vision in vertebrates. Eye, 12 (Pt 3b), 541–547.

Boyd, R., & Silk, J. B. (2000). How humans evolved, 2d ed. New York: W. W. Norton.

Bradbury, E. J., Moon, L. D., Popat, R. J., King, V. R., Bennett, G. S., Patel, P. N., et al. (2002). Chondroitinase ABC promotes functional recovery after spinal cord injury. Nature, 416(6881), 636–640.

Brakefield, P. M., Gates, J., Keys, D., Kesbeke, F., Wijngaarden, P. J., Monteiro, A., et al. (1996). Development, plasticity, and evolution of butterfly eyespot patterns. Nature, 384, 236–243.

Brockes, J. P. (1997). Amphibian limb regeneration: rebuilding a complex structure. Science, 276(5309), 81–87.

Brown, A., Yates, P. A., Burrola, P., Ortuno, D., Vaidya, A., Jessell, T. M., et al. (2000). Topographic mapping from the retina to the midbrain is controlled by relative but not absolute levels of EphA receptor signaling. Cell, 102(1), 77–88.

Brown, R. W. (1957). Linguistic determinism and the part of speech. J Abnorm & Soc Psychol, 55, 1–5.

Brunet, M., Guy, F., Pilbeam, D., Mackaye, H. T., Likius, A., Ahounta, D., et al. (2002). A new hominid from the Upper Miocene of Chad, Central Africa. Nature, 418(6894), 145–151.

Buchsbaum, M. S., Tang, C. Y., Peled, S., Gudbjartsson, H., Lu, D., Hazlett, E. A., et al. (1998). MRI white matter diffusion anisotropy and PET metabolic rate in schizophrenia. Neuroreport, 9(3), 425–430.

Buckner, R. L., Kelley, W. M., & Petersen, S. E. (1999). Frontal cortex contributes to human memory formation. Nat Neurosci, 2(4), 311–314.

Bullock, T. H., Moore, J. K., & Fields, R. D. (1984). Evolution of myelin sheaths: both lamprey and hagfish lack myelin. Neurosci Lett, 48(2), 145–148.

Butler, A. B., & Hodos, W. (1996). Comparative vertebrate neuroanatomy: evolution and adaptation. New York: Wiley-Liss.

Buxhoeveden, D. P., Switala, A. E., Litaker, M., Roy, E., & Casanova, M. F. (2001). Lateralization of minicolumns in human planum temporale is absent in non-human primate cortex. Brain Behav Evol, 57(6), 349–358.

Cacabelos, R. (2002). Pharmacogenomics for the treatment of dementia. Ann Med, 34(5), 357–379.

Caldwell, C. A., & Whiten, A. (2002). Evolutionary perspectives on imitation: is a comparative psychology of social learning possible? Anim Cogn, 5(4), 193–208.

Call, J., & Tomasello, M. (1999). A nonverbal false belief task: the performance of children and great apes. Child Dev, 70(2), 381–395.

Cantalupo, C., & Hopkins, W. D. (2001). Asymmetric Broca's area in great apes. Nature, 414(6863), 505.

Carey, S. (1978). The child as word-learner. In M. Halle, J. Bresnan, & G. A. Miller (eds.), Linguistic theory and psychological reality. Cambridge: MIT Press.

Carroll, S. B., Grenier, J. K., & Weatherbee, S. D. (2001). From DNA to diversity: molecular genetics and the evolution of animal design. Oxford and Malden, Mass.: Blackwell Science.

Carruthers, P., & Boucher, J. (1998). Language and thought: interdisciplinary themes. Cambridge and New York: Cambridge University Press.

Carugo, O., & Pongor, S. (2002). The evolution of structural databases. Trends Biotechnol, 20(12), 498–501.

Caspi, A., McClay, J., Moffitt, T. E., Mill, J., Martin, J., Craig, I. W., et al. (2002). Role of genotype in the cycle of violence in maltreated children. Science, 297(5582), 851–854.

Castren, E., Zafra, F., Thoenen, H., & Lindholm, D. (1992). Light regulates expression of brain-derived neurotrophic factor mRNA in rat visual cortex. Proc Natl Acad Sci USA, 89(20), 9444–9448.

Cebria, F., Kobayashi, C., Umesono, Y., Nakazawa, M., Mineta, K., Ikeo, K., et al. (2002). FGFR-related gene nou-darake restricts brain tissues to the head region of planarians. Nature, 419(6907), 620–624.

Cebria, F., Nakazawa, M., Mineta, K., Ikeo, K., Gojobori, T., & Agata, K. (2002). Dissecting planarian central nervous system regeneration by the expression of neural-specific genes. Dev Growth Differ, 44(2), 135–146.

Chargaff, E. (1950). Chemical specificity of nucleic acids and mechanism of their enzymatic degradation. Experientia, 6, 201–209.

Chen, F. C., & Li, W. H. (2001). Genomic divergences between humans and other hominoids and the effective population size of the common ancestor of humans and chimpanzees. Am J Hum Genet, 68(2), 444–456.

Cheney, D. L., & Seyfarth, R. M. (1990). How monkeys see the world: inside the mind of another species. Chicago: University of Chicago Press.

Chenn, A., & Walsh, C. A. (2002). Regulation of cerebral cortical size by control of cell cycle exit in neural precursors. Science, 297(5580), 365–369.

Chew, S. J., Mello, C., Nottebohm, F., Jarvis, E., & Vicario, D. S. (1995). Decrements in auditory responses to a repeated conspecific song are long-lasting and require two periods of protein synthesis in the songbird forebrain. Proc Natl Acad Sci USA, 92(8), 3406–3410.

Chomsky, N. A. (1965). Aspects of a theory of syntax. Cambridge: MIT Press.

___. (1975). Reflections on language. New York: Pantheon.

___. (2000). New horizons in the study of language and mind. Cambridge and New York: Cambridge University Press.

Clancy, B., Darlington, R. B., & Finlay, B. L. (2001). Translating developmental time across mammalian species. Neurosci, 105(1), 7–17.

Clandinin, T. R., & Zipursky, S. L. (2002). Making connections in the fly visual system. Neuron, 35(5), 827–841.

Clifton, R. K., Perris, E. E., & McCall, D. D. (1999). Does reaching in the dark for unseen objects reflect representation in infants? Infant Behav & Dev, 22(3), 297–302.

Clifton, R. K., Rochat, P., Litovsky, R. Y., & Perris, E. E. (1991). Object representation guides infants' reaching in the dark. J Exp Psychol: Hum Percept & Perform, 17(2), 323–329.

Cohen, L. B., & Marks, K. S. (2002). How infants process addition and subtraction events. Dev Sci, 5(2), 186–201.

Coppinger, R., & Coppinger, L. (2001). Dogs: a startling new understanding of canine origin, behavior, and evolution. New York: Scribner.

Corballis, M. C. (1992). On the evolution of language and generativity. Cognition, 44(3), 197–126.

Coren, S. (1994). The intelligence of dogs: canine consciousness and capabilities. New York: Free Press.

Costa, R. M., Federov, N. B., Kogan, J. H., Murphy, G. G., Stern, J., Ohno, M., et al. (2002). Mechanism for the learning deficits in a mouse model of neurofibromatosis type 1. Nature, 415(6871), 526–530.

Costagli, A., Kapsimali, M., Wilson, S. W., & Mione, M. (2002). Conserved and divergent patterns of Reelin expression in the zebrafish central nervous system. J Comp Neurol, 450(1), 73–93.

Crawley, J. N. (2000). What's wrong with my mouse? behavioral phenotyping of transgenic and knockout mice. New York: Wiley-Liss.

Crick, F. (1993). The astonishing hypothesis: the scientific search for the soul. New York: Scribner.

Crossley, P. H., Martinez, S., Ohkubo, Y., & Rubenstein, J. L. (2001). Coordinate expression of Fgf8, Otx2, Bmp4, and Shh in the rostral prosencephalon during development of the telencephalic and optic vesicles. Neurosci, 108(2), 183–206.

Crosson, B. (1992). Subcortical functions in language and memory. New York: Guilford.

Crowley, J. C., & Katz, L. C. (1999). Development of ocular dominance columns in the absence of retinal input. Nature Neurosci, 2(12), 1125–1130.

Cruz, Y. P. (1997). Mammals. In S. F. Gilbert & A. M. Raunio (eds.), Embryology: constructing the organism, pp. 459–489. Sunderland, Mass.: Sinauer Associates.

Damasio, A. R. (1994). Descartes' error: emotion, reason, and the human brain. New York: Putnam.

Damasio, H., Grabowski, T. J., Tranel, D., Hichwa, R. D., & Damasio, A. R. (1996). A neural basis for lexical retrieval. Nature, 380(6574), 499–505.

Darwin, C. (1874). The descent of man and selection in relation to sex, 2d ed. New York: D. Appleton.

Davidoff, J., Davies, I., & Roberson, D. (1999). Colour categories in a stone-age tribe. Nature, 398(6724), 203–204.

Davidson, E. H. (2001). Genomic regulatory systems: development and evolution. San Diego: Academic Press.

Davis, H. P., & Squire, L. R. (1984). Protein synthesis and memory: a review. Psychol Bull, 96(3), 518–559.

Davis, J. (1997). Mapping the mind: the secrets of the human brain and how it works. Secaucus, N.J.: Carol Pub. Group.

Daw, N. W. (1994). Mechanisms of plasticity in the visual cortex. The Friedenwald Lecture. Invest Ophthalmol Vis Sci, 35(13), 4168–4179.

de Bode, S., & Curtiss, S. (2000). Language after hemispherectomy. Brain Cogn, 43(1–3), 135–138.

de Duve, C. (1995). The beginnings of life on earth. Am Scient, 83, 428–437.

de Winter, W., & Oxnard, C. E. (2001). Evolutionary radiations and convergences in the structural organization of mammalian brains. Nature, 409(6821), 710–714.

Deacon, T. W. (1997). The symbolic species: the co-evolution of language and the brain. New York: W. W. Norton.

———. (2000). Evolutionary perspectives on language and brain plasticity. J Commun Disord, 33(4), 273–290.

DeCasper, A. J., & Spence, M. J. (1986). Prenatal maternal speech influences newborns' perception of speech sounds. Infant Behav & Dev, 9(2), 133–150.

DeFries, J. C., Gervais, M. C., & Thomas, E. A. (1978). Response to 30 generations of selection for open-field activity in laboratory mice. Behav Genet, 8(1), 3–13.

Dehaene, S. (1997). The number sense: how the mind creates mathematics. New York: Oxford University Press.

Diamond, J. M. (1992). The third chimpanzee: the evolution and future of the human animal. New York: HarperCollins.

Dickens, W. T., & Flynn, J. R. (2001). Heritability estimates versus large environmental effects: the IQ paradox resolved. Psychol Rev, 108(2), 346–369.

Dietrich, A., & Been, W. (2001). Memory and DNA. J Theor Biol, 208(2), 145–149.

Dromi, E. (1987). Early lexical development. New York: Cambridge University Press.

Dronkers, N. F. (2000). The pursuit of brain-language relationships. Brain Lang, 71(1), 59–61.

Dubnau, J., Chiang, A. S., Grady, L., Barditch, J., Gossweiler, S., McNeil, J., et al. (2003). The staufen/pumilio pathway is involved in Drosophila long-term memory. Curr Biol, 13(4), 286–296.

Dubnau, J., Chiang, A. S., & Tully, T. (2003). Neural substrates of memory: from synapse to system. J Neurobiol, 54(1), 238–253.

Dunbar, R. I. M. (1996). Grooming, gossip, and the evolution of language. Cambridge: Harvard University Press.

Dunlop, S. A., Lund, R. D., & Beazley, L. D. (1996). Segregation of optic input in a three-eyed mammal. Exp Neurol, 137(2), 294–298.

Dunlop, S. A., Rodger, J., King, C., Stirling, R. V., Tee, L., Ziman, M., et al. (2002). Molecular events during optic nerve regeneration. Int J Dev Neurosci, 19(1), 693.

Dunlop, S. A., Tran, N., Tee, L. B., Papadimitriou, J., & Beazley, L. D. (2000). Retinal projections throughout optic nerve regeneration in the ornate dragon lizard, Ctenophorus ornatus. J Comp Neurol, 416(2), 188–200.

Ebersberger, I., Metzler, D., Schwarz, C., & Paabo, S. (2002). Genomewide comparison of DNA sequences between humans and chimpanzees. Am J Hum Genet, 70(6), 1490–1497.

Eddy, S. R. (2001). Non-coding RNA genes and the modern RNA world. Nat Rev Genet, 2(12), 919–929.

Egan, M. F., Kojima, M., Callicott, J. H., Goldberg, T. E., Kolachana, B. S., Bertolino, A., et al. (2003). The BDNF val66met polymorphism affects activity-dependent secretion of BDNF and human memory and hippocampal function. Cell, 112(2), 257–269.

Ehrlich, P. R. (2000). Human natures: genes, cultures, and the human prospect. Washington, D.C.: Island Press/Shearwater Books.

Ellis, N. C., & Hennelly, R. A. (1980). A bilingual word-length effect: implications for intelligence testing and the relative ease of mental calculation in Welsh and English. Brit J Psychol, 71(1), 43–51.

Embick, D., Marantz, A., Miyashita, Y., O'Neil, W., & Sakai, K. L. (2000). A syntactic specialization for Broca's area. Proc Natl Acad Sci USA, 97(11), 6150–6154.

Emlen, S. T. (1975). The stellar-orientation system of a migratory bird. Sci Am, 233(2), 102–111.

Emmons, S. W., & Lipton, J. (2003). Genetic basis of male sexual behavior. J Neurobiol, 54(1), 93–110.

Enard, W., Przeworski, M., Fisher, S. E., Lai, C. S., Wiebe, V., Kitano, T., et al. (2002). Molecular evolution of FOXP2, a gene involved in speech and language. Nature, 418(6900), 869–872.

Evans, W. E., & Relling, M. V. (1999). Pharmacogenomics: translating functional genomics into rational therapeutics. Science, 286(5439), 487–491.

Fantz, R. L. (1961). The origin of form perception. Sci Am, 204, 66–72.

Farroni, T., Csibra, G., Simion, F., & Johnson, M. H. (2002). Eye contact detection in humans from birth. Proc Natl Acad Sci USA, 99(14), 9602–9605.

Fatemi, S. H. (2002). The role of Reelin in pathology of autism. Mol Psych, 7(9), 919–920.

Fauconnier, G., & Turner, M. (2002). The way we think: conceptual blending and the mind's hidden complexities. New York: Basic Books.

Feldman, J. A., & Ballard, D. H. (1982). Connectionist models and their properties. Cogn Sci, 6, 205–254.

Fenson, L., Dale, P. S., Reznick, J. S., Bates, E., Thal, D. J., & Pethick, S. J. (1994). Variability in early communicative development. Monogr Soc Res Child Dev, 59(5), 1–173; discussion 174–185.

Fentress, J. C. (1973). Development of grooming in mice with amputated forelimbs. Science, 179(74), 704–705.

Fields, R. D., Eshete, F., Dudek, S., Ozsarac, N., & Stevens, B. (2001). Regulation of gene expression by action potentials: dependence on complexity in cellular information processing. Novartis Found Symp, 239, 160–172; discussion 172–166, 234–240.

Fields, R. D., Eshete, F., Stevens, B., & Itoh, K. (1997). Action potential-dependent regulation of gene expression: temporal specificity in ca2+, cAMP-responsive element binding proteins, and mitogen-activated protein kinase signaling. J Neurosci, 17(19), 7252–7266.

Finlay, B. L., Darlington, R. B., & Nicastro, N. (2001). Developmental structure in brain evolution. Behav Brain Sci, 24(2), 263–278; discussion 278–308.

Fisher, S. E., & DeFries, J. C. (2002). Developmental dyslexia: genetic dissection of a complex cognitive trait. Nat Rev Neurosci, 3(10), 767–780.

Fisher, S. E., Vargha-Khadem, F., Watkins, K. E., Monaco, A. P., & Pembrey, M. E. (1998). Localisation of a gene implicated in a severe speech and language disorder. Nat Genet, 18(2), 168–170. Erratum in Nat Genet 18(3), March 1998, 298.

Fitch, W. T. (2000). The evolution of speech: a comparative review. Trends Cogn Sci, 4(7), 258–267.

Flexner, L. B., Flexner, J. B., & Stellar, E. (1965). Memory and cerebral protein synthesis in mice as affected by graded amounts of puromycin. Exp Neurol, 13(3), 264–272.

Flores, G. V., Duan, H., Yan, H., Nagaraj, R., Fu, W., Zou, Y., et al. (2000). Combinatorial signaling in the specification of unique cell fates. Cell, 103(1), 75–85.

Flynn, J. R. (1999). Searching for justice: the discovery of IQ gains over time. Am Psychol, 54(1), 5–20.

Fodor, J. A. (1975). The language of thought. New York: T. Y. Crowell.

Frank, M., & Kemler, R. (2002). Protocadherins. Curr Opin Cell Biol, 14(5), 557–562.

Friederici, A. D. (2002). Towards a neural basis of auditory sentence processing. Trends Cogn Sci, 6(2), 78–84.

Fukuchi-Shimogori, T., & Grove, E. A. (2001). Neocortex patterning by the secreted signaling molecule FGF8. Science, 294(5544), 1071–1074.

Fukuyama, F. (2002). Our posthuman future: consequences of the biotechnology revolution. New York: Farrar, Straus and Giroux.

Furlow, J. D., & Brown, D. D. (1999). In vitro and in vivo analysis of the regulation of a transcription factor gene by thyroid hormone during Xenopus laevis metamorphosis. Mol Endocrinol, 13(12), 2076–2089.

Fuster, J. M. (2000). Memory networks in the prefrontal cortex. Prog Brain Res, 122, 309–316.

Gainetdinov, R. R., Wetsel, W. C., Jones, S. R., Levin, E. D., Jaber, M., & Caron, M. G. (1999). Role of serotonin in the paradoxical calming effect of psychostimulants on hyperactivity. Science, 283(5400), 397–401.

Galli, L., & Maffei, L. (1988). Spontaneous impulse activity of rat retinal ganglion cells in prenatal life. Science, 242(4875), 90–91.

Gallistel, C. R. (1990). The organization of learning. Cambridge: MIT Press.

Gallistel, C. R. (2002). The principle of adaptive specialization as it applies to learning and memory. In R. H. Kluwe, G. Lüer, & F. Rösler (eds.), Principles of human learning and memory, pp. 250–280. Basel: Birkenaeuser.

Gallistel, C. R., Brown, A., Carey, S., Gelman, R., & Keil, F. (1991). Lessons from animal learning. In R. Gelman & S. Carey (eds.), The epigenesis of mind, pp. 3–37. Hillsdale, N.J.: Lawrence Erlbaum Associates.

Gannon, P. J., Holloway, R. L., Broadfield, D. C., & Braun, A. R. (1998). Asymmetry of chimpanzee planum temporale: humanlike pattern of Wernicke's brain language area homolog. Science, 279(5348), 220–222.

Garaschuk, O., Linn, J., Eilers, J., & Konnerth, A. (2000). Large-scale oscillatory calcium waves in the immature cortex. Nat Neurosci, 3(5), 452–459.

Garcia-Fernandez, J., & Holland, P. W. (1994). Archetypal organization of the amphioxus Hox gene cluster. Nature, 370(6490), 563–566.

Garrod, A. E. (1923). Inborn errors of metabolism, 2d ed. London: H. Frowde and Hodder & Stoughton.

Gasking , E. B. (1959). Why was Mendel's work ignored? J Hist Ideas, 20, 60–84.

Gazzaniga, M. S. (1998). The split brain revisited. Sci Am, 279(1), 50–55.

Gehring, W. J. (1998). Master control genes in development and evolution: the homeobox story. New Haven: Yale University Press.

____. (2002). The genetic control of eye development and its implications for the evolution of the various eye-types. Int J Dev Biol, 46(1), 65–73.

Gerhart, J., & Kirschner, M. (1997). Cells, embryos, and evolution. Cambridge: Blackwell Science.

Geschwind, N., & Levitsky, W. (1968). Human brain: left-right asymmetries in temporal speech region. Science, 161(837), 186–187.

Gilbert, S. F. (2000). Developmental biology, 6th ed. Sunderland, Mass.: Sinauer Associates.

Gleitman, L. R., Gleitman, H., Miller, C., & Ostrin, R. (1996). Similar, and similar concepts. Cognition, 58(3), 321–376.

Gödecke, I., & Bonhoeffer, T. (1996). Development of identical orientation maps for two eyes without common visual experience. Nature, 379, 251–254.

Godwin, J., Luckenbach, J. A., & Borski, R. J. (2003). Ecology meets endocrinology: environmental sex determination in fishes. Evol Dev, 5(1), 40–49.

Goho, A. M. (2003). Life made to order. Tech Rev (April), 50–57.

Goldman-Rakic, P. S., Bourgeois, J. P., & Rakic, P. (1997). Synaptic substrate of cognitive development. In N. A. Krasnegor & G. R. Lyon & P. S. Goldman-Rakic (eds.), Development of the prefrontal cortex: evolution, neurobiology, and behavior, pp. 27–47. Baltimore: Paul H. Brookes.

Golling, G., Amsterdam, A., Sun, Z., Antonelli, M., Maldonado, E., Chen, W., et al. (2002). Insertional mutagenesis in zebrafish rapidly identifies genes essential for early vertebrate development. Nat Genet, 31(2), 135–140.

Gomez, R. L., & Gerken, L.-A. (1999). Artificial grammar learning by 1-year-olds leads to specific and abstract knowledge. Cognition, 70(1), 109–135.

Goodglass, H. (1993). Understanding aphasia. San Diego: Academic Press.

Goodhill, G. J. (1998). Mathematical guidance for axons. Trends Neurosci, 21(6), 226–231.

Goodhill, G. J., & Richards, L. J. (1999). Retinotectal maps: molecules, models and misplaced data. Trends Neurosci, 22(12), 529–534.

Goodman, C. S. (1978). Isogenic grasshoppers: genetic variability in the morphology of identified neurons. J Comp Neurol, 182(4), 681–705.

Gopnik, M., & Crago, M. B. (1991). Familial aggregation of a developmental language disorder. Cognition, 39(1), 1–50.

Goren, C. C., Sarty, M., & Wu, P. Y. (1975). Visual following and pattern discrimination of face-like stimuli by newborn infants. Pediatrics, 56(4), 544–549.

Gould, E., Reeves, A. J., Graziano, M. S., & Gross, C. G. (1999). Neurogenesis in the neocortex of adult primates. Science, 286(5439), 548–552.

Gould, S. J. (1977). Ontogeny and phylogeny. Cambridge: Belknap Press of Harvard University Press.

___. (1979). Panselectionist pitfalls in Parker & Gibson's model of the evolution of intelligence. Behav and Brain Sci, 2, 385–386.

Gray, J. (2002). Straw dogs: thoughts on humans and other animals. London: Granta.

Greer, J. M., & Capecchi, M. R. (2002). Hoxb8 is required for normal grooming behavior in mice. Neuron, 33(1), 23–34.

Gregersen, P. K., Kowalsky, E., Kohn, N., & Marvin, E. W. (2001). Early childhood music education and predisposition to absolute pitch: teasing apart genes and environment. Am J Med Genet, 98(3), 280–282.

Grimsby, J., Toth, M., Chen, K., Kumazawa, T., Klaidman, L., Adams, J. D., et al. (1997). Increased stress response and beta-phenylethylamine in MAOB-deficient mice. Nat Genet, 17(2), 206–210.

Gross, C., Zhuang, X., Stark, K., Ramboz, S., Oosting, R., Kirby, L., et al. (2002). Serotonin 1A receptor acts during development to establish normal anxiety-like behaviour in the adult. Nature, 416(6879), 396–400.

Gumperz, J. J., & Levinson, S. C. (1996). Rethinking linguistic relativity. Cambridge and New York: Cambridge University Press.

Gustafsson, T., & Kraus, W. E. (2001). Exercise-induced angiogenesis-related growth and transcription factors in skeletal muscle, and their modification in muscle pathology. Front Biosci, 6, D75–89.

Hall, J. C. (1994). The mating of a fly. Science, 264(5166), 1702–1714.

Hall, J. M., Lee, M. K., Newman, B., Morrow, J. E., Anderson, L. A., Huey, B., et al. (1990). Linkage of early-onset familial breast cancer to chromosome 17q21. Science, 250(4988), 1684–1689.

Hall, Z. W. (1992). The cells of the nervous system. In Z. W. Hall (ed.), An introduction to molecular neurobiology, pp. 1–29. Sunderland, Mass.: Sinauer Associates.

Hannan, A. J., Blakemore, C., Katsnelson, A., Vitalis, T., Huber, K. M., Bear, M., et al. (2001). PLC-beta1, activated via mGluRs, mediates activity-dependent differentiation in cerebral cortex. Nat Neurosci, 4(3), 282–288.

Hare, B., Brown, M., Williamson, C., & Tomasello, M. (2002). The domestication of social cognition in dogs. Science, 298(5598), 1634–1636.

Hariri, A. R., Mattay, V. S., Tessitore, A., Kolachana, B., Fera, F., Goldman, D., et al. (2002). Serotonin transporter genetic variation and the response of the human amygdala. Science, 297(5580), 400–403.

Harris, B. (1979). Whatever happened to little Albert? Am Psychol, 34(2), 151–160.

Harris, J., Honigberg, L., Robinson, N., & Kenyon, C. (1996). Neuronal cell migration in C. elegans: regulation of Hox gene expression and cell position. Development, 122(10), 3117–3131.

Harris, W. A. (1986). Homing behaviour of axons in the embryonic vertebrate brain. Nature, 320(6059), 266–269.

Harris, W. A., Holt, C. E., & Bonhoeffer, F. (1987). Retinal axons with and without their somata, growing to and arborizing in the tectum of Xenopus embryos: a time-lapse video study of single fibres in vivo. Development, 101(1), 123–133.

Harris-Warrick, R. M. (2000). Ion channels and receptors: molecular targets for behavioral evolution. J Comp Physiol [A], 186(7–8), 605–616.

Hauser, M. D. (2002). Ontogeny of tool use in cottontop tamarins, Saguinus oedipus: innate recognition of functionally relevant features. Anim Behav, 64, 299–311.

Hauser, M. D., Chomsky, N., & Fitch, W. T. (2002). The faculty of language: what is it, who has it, and how did it evolve? Science, 298(5598), 1569–1579.

Hauser, M. D., Weiss, D., & Marcus, G. (2002). Rule learning by cotton-top tamarins. Cognition, 86(1), B15.

Heath, S. B. (1983). Ways with words. New York: Cambridge University Press.

Hebb, D. O. (1947). The effect of early experience on problem-solving at maturity. Am Psychol, 2, 306–307.

Hedges, L. V., & Nowell, A. (1995). Sex differences in mental test scores, variability, and numbers of high-scoring individuals. Science, 269(5220), 41–45.

Hershey, A. D., & Chase, M. (1952). Independent functions of viral proteins and nucleic acid in growth of bacteriophage. J Gen Physiol, 36, 39–56.

Hibbard, E. (1965). Orientation and directed growth of Mauther's cell axons from duplicated vestibular nerve roots. Exp Neurol, 13, 289–301.

Hilschmann, N., Barnikol, H. U., Barnikol-Watanabe, S., Gotz, H., Kratzin, H., & Thinnes, F. P. (2001). The immunoglobulin-like genetic predetermination of the brain: the protocadherins, blueprint of the neuronal network. Naturwissenschaften, 88(1), 2–12.

Hirotsune, S., Yoshida, N., Chen, A., Garrett, L., Sugiyama, F., Takahashi, S., et al. (2003). An expressed pseudogene regulates the messenger-RNA stability of its homologous coding gene. Nature, 423(6935), 91–96.

Hobert, O. (2003). Behavioral plasticity in C. elegans: paradigms, circuits, genes. J Neurobiol, 54(1), 203–223.

Hochedlinger, K., & Jaenisch, R. (2002). Nuclear transplantation: lessons from frogs and mice. Curr Opin Cell Biol, 14(6), 741–748.

Hofmann, H. A. (2003). Functional genomics of neural and behavioral plasticity. J Neurobiol, 54(1), 272–282.

Hogenesch, J. B., Ching, K. A., Batalov, S., Su, A. I., Walker, J. R., Zhou, Y., et al. (2001). A comparison of the Celera and Ensembl predicted gene sets reveals little overlap in novel genes. Cell, 106(4), 413–415.

Holland, H. D. (1997). Evidence for life on earth more than 3850 million years ago. Science, 275(5296), 38–39.

Holliday, R. (1999). Is there an epigenetic component in long-term memory? J Theor Biol, 200(3), 339–341.

Hoosain, R., & Salili, F. (1987). Language differences in pronunciation speed for numbers, digit span, and mathematical ability. Psychologia: Internatl J Psychol in the Orient, 30(1), 34–38.

Hopkin, K. (2001). The post-genome project. Sci Am, 285(2), 16.

Horn, G., & McCabe, B. J. (1984). Predispositions and preferences: effects of imprinting of lesions to the chick brain. Anim Behav, 32, 288–292.

Hsiao, L. L., Dangond, F., Yoshida, T., Hong, R., Jensen, R. V., Misra, J., et al. (2001). A compendium of gene expression in normal human tissues. Physiol Genomics, 7(2), 97–104.

Hubel, D. H. (1988). Eye, brain, and vision. New York: Scientific American Library.

Hubel, D. H., & Wiesel, T. N. (1962). Receptive fields, binocular interaction and functional architecture in the cat's visual cortex. J Physiol, 160, 106–154.

Huffman, K. J., Molnar, Z., Van Dellen, A., Kahn, D. M., Blakemore, C., & Krubitzer, L. (1999). Formation of cortical fields on a reduced cortical sheet. J Neurosci, 19(22), 9939–9952.

Hunt, K. K., & Vorburger, S. A. (2002). Gene therapy. Hurdles and hopes for cancer treatment. Science, 297(5580), 415–416.

Husi, H., & Grant, S. G. (2001). Proteomics of the nervous system. Trends Neurosci, 24(5), 259–266.

Hutchinson, E. (2001). Working towards tailored therapy for cancer. Lancet, 357(9267), 1508.

Huttenlocher, J. (1998). Language input and language growth. Prev Med, 27(2), 195–199.

Huttenlocher, P. R. (1990). Morphometric study of human cerebral cortex development. Neuropsychologia, 28(6), 517–527.

Ikonomidou, C., Bittigau, P., Koch, C., Genz, K., Hoerster, F., Felderhoff-Mueser, U., et al. (2001). Neurotransmitters and apoptosis in the developing brain. Biochem Pharmacol, 62(4), 401–405.

International Human Genome Sequencing Consortium. (2001). Initial sequencing and analysis of the human genome. Nature, 409, 860–921.

Ioshikhes, I. P., & Zhang, M. Q. (2000). Large-scale human promoter mapping using CpG islands. Nat Genet, 26(1), 61–63.

Isacson, O., & Deacon, T. (1997). Neural transplantation studies reveal the brain's capacity for continuous reconstruction. Trends Neurosci, 20(10), 477–482.

Jaaro, H., Beck, G., Conticello, S. G., & Fainzilber, M. (2001). Evolving better brains: a need for neurotrophins? Trends Neurosci, 24(2), 79–85.

Jackendoff, R. (2002). Foundations of language: brain, meaning, grammar, evolution. Oxford and New York: Oxford University Press.

Jacob, F., & Monod, J. (1961). On the regulation of gene activity. Cold Spring Harbor Symposium on Quantitative Biology, 26, 193–211.

Jameson, K. A., Highnote, S. M., & Wasserman, L. M. (2001). Richer color experience in observers with multiple photopigment opsin genes. Psychon Bull Rev, 8(2), 244–261.

Jeeves, M. A., & Temple, C. M. (1987). A further study of language function in callosal agenesis. Brain Lang, 32(2), 325–335.

Jegla, T., & Salkoff, L. (1994). Molecular evolution of K+ channels in primitive eukaryotes. Soc Gen Physiol Ser, 49, 213–222.

Jerison, H. J. (1979). The evolution of diversity in brain size. In M. E. Hahn, C. Jensen, & B. C. Dudek (eds.), Development and evolution of brain size: behavioral implications, pp. 29–57. New York: Academic Press.

Johnson, J. S., & Newport, E. L. (1989). Critical period effects in second language learning: the influence of maturational state on the acquisition of English as second language. Cogn Psychol, 21, 60–99.

Johnson, M. H. (1997). Developmental cognitive neuroscience. Oxford: Basil Blackwell.

Johnson, M. H., Bolhuis, J. J., & Horn, G. (1985). Interaction between acquired preferences and developing predispositions during imprinting. Anim Behav, 33, 1000–1006.

Johnson, M. H., & Morton, J. (1991). Biology and cognitive development: the case of face recognition. Oxford and Cambridge: Basil Blackwell.

Johnson, S. C., & Carey, S. (1998). Knowledge enrichment and conceptual change in folkbiology: evidence from Williams syndrome. Cogn Psychol, 37(2), 156–200.

Judson, H. F. (1979). The eighth day of creation: makers of the revolution in biology. New York: Simon & Schuster.

Kaan, E., & Swaab, T. Y. (2002). The brain circuitry of syntactic comprehension. Trends Cogn Sci, 6(8), 350–356.

Kaas, J. H. (1987). The organization of neocortex in mammals: implications for theories of brain function. Annu Rev Psychol, 38, 129–151.

___. (2002). Sensory loss and cortical reorganization in mature primates. Prog Brain Res, 138, 167–176.

Kaczmarek, L. (1993). Molecular biology of vertebrate learning: is c-fos a new beginning? J Neurosci Res, 34(4), 377–381.

___. (2000). Gene expression in learning processes. Acta Neurobiol Exp (Warsz), 60(3), 419–424.

Kaczmarek, L., Zangenehpour, S., & Chaudhuri, A. (1999). Sensory regulation of immediate-early genes c-fos and zif268 in monkey visual cortex at birth and throughout the critical period. Cereb Cortex, 9(2), 179–187.

Kandel, E. R. (1979). Behavioral biology of Aplysia: a contribution to the comparative study of opisthobranch molluscs. San Francisco: W. H. Freeman.

___. (2001). The molecular biology of memory storage: a dialogue between genes and synapses. Science, 294(5544), 1030–1038.

Kandel, E. R., & O'Dell, T. J. (1992). Are adult learning mechanisms also used for development? Science, 258(5080), 243–245.

Kandel, E. R., Schwartz, J. H., & Jessell, T. M. (2000). Principles of neural science, 4th ed. New York: McGraw-Hill Health Professions Division.

Karmiloff, K., & Karmiloff-Smith, A. (2001). Pathways to language: from fetus to adolescent. Cambridge: Harvard University Press.

Karmiloff-Smith, A. (1998). Development itself is the key to understanding developmental disorders. Trends Cogn Sci, 2, 389–398.

Kaschube, M., Wolf, F., Geisel, T., & Lowel, S. (2002). Genetic influence on quantitative features of neocortical architecture. J Neurosci, 22(16), 7206–7217.

Katz, L. C., & Shatz, C. J. (1996). Synaptic activity and the construction of cortical circuits. Science, 274, 1133–1138.

Kimble, J., & Austin, J. (1989). Genetic control of cellular interactions in Caenorhabditis elegans development. Ciba Found Symp, 144, 212–220; discussion 221–226, 290–295.

King, M. C., & Wilson, A. C. (1975). Evolution at two levels in humans and chimpanzees. Science, 188(4184), 107–116.

Klein, R. G., & Edgar, B. (2002). The dawn of human culture. New York: Wiley.

Klintsova, A. Y., & Greenough, W. T. (1999). Synaptic plasticity in cortical systems. Curr Opin Neurobiol, 9(2), 203–208.

Klug, A. (1974). Rosalind Franklin and the double helix. Nature, 248(451), 787–788.

Knudsen, E. I., & Knudsen, P. F. (1990). Sensitive and critical periods for visual calibration of sound localization by barn owls. J Neurosci, 10(1), 222–232.

Koch, C., & Segev, I. (2000). The role of single neurons in information processing. Nat Neurosci, 3 Suppl, 1171–1177.

Koelsch, S., Gunter, T. C., von Cramon, D. Y., Zysset, S., Lohmann, G., & Friederici, A. D. (2002). Bach speaks: a cortical "language-network" serves the processing of music. Neuroimage, 17(2), 956–966.

Komiyama, T., Johnson, W. A., Luo, L., & Jefferis, G. S. (2003). From lineage to wiring specificity: POU domain transcription factors control precise connections of Drosophila olfactory projection neurons. Cell, 112(2), 157–167.

Kooy, R. F. (2003). Of mice and the fragile X syndrome. Trends Genet, 19(3), 148–154.

Korenberg, J. R., Chen, X. N., Hirota, H., Lai, Z., Bellugi, U., Burian, D., et al. (2000). VI. Genome structure and cognitive map of Williams syndrome. J Cogn Neurosci, 12 Suppl 1, 89–107.

Kranzler, J. H., Rosenbloom, A. L., Martinez, V., & Guevara-Aguirre, J. (1998). Normal intelligence with severe insulin-like growth factor I deficiency due to growth hormone receptor deficiency: a controlled study in a genetically homogeneous population. J Clin Endocrinol Metab, 83(6), 1953–1958.

Kroodsma, D. E. (1984). Songs of the alder flycatcher (Empidonax alnorum) and willow flycatcher (Empidonax trailhi) are innate. Auk, 10, 13–24.

Krubitzer, L., & Huffman, K. J. (2000). Arealization of the neocortex in mammals: genetic and epigenetic contributions to the phenotype. Brain Behav Evol, 55(6), 322–335.

Krubitzer, L. A. (2000). How does evolution build a complex brain? Novartis Found Symp, 228, 206–220; discussion 220–226.

Kujala, T., Alho, K., & Naatanen, R. (2000). Cross-modal reorganization of human cortical functions. Trends Neurosci, 23(3), 115–120.

Kullander, K., Butt, S. J., Lebret, J. M., Lundfald, L., Restrepo, C. E., Rydstrom, A., et al. (2003). Role of EphA4 and EphrinB3 in local neuronal circuits that control walking. Science, 299 (5614), 1889–1892.

Kumar, A., & Cook, I. A. (2002). White matter injury, neural connectivity and the pathophysiology of psychiatric disorders. Dev Neurosci, 24(4), 255–261.

Lacalli, T. C. (2001). New perspectives on the evolution of protochordate sensory and locomotory systems, and the origin of brains and heads. Philos Trans R Soc Lond B Biol Sci, 356(1414), 1565–1572.

Lai, C. S., Fisher, S. E., Hurst, J. A., Vargha-Khadem, F., & Monaco, A. P. (2001). A forkhead-domain gene is mutated in a severe speech and language disorder. Nature, 413(6855), 519–523.

Lander, E. S., & Schork, N. J. (1994). Genetic dissection of complex traits. Science, 265(5181), 2037–2048.

Larsen, B. H., Vestergaard, K. S., & Hogan, J. A. (2000). Development of dust-bathing behavior sequences in the domestic fowl: the significance of functional experience. Dev Psychobiol, 37(1), 5–12.

Law, M. I., & Constantine-Paton, M. (1981). Anatomy and physiology of experimentally produced striped tecta. J Neurosci, 1(7), 741–759.

Lawn, I. D., Mackie, G. O., & Silver, G. (1981). Conduction system in a sponge. Science, 211(4487), 1169–1171.

Lebon, I. (2001). Preliminary results of the human genome project: a scientific paper in two parts. McSweeney's.

Ledoux, J. E. (1996). The emotional brain: the mysterious underpinnings of emotional life. New York: Simon & Schuster.

Lee, H. C., Ladd, C., Bourke, M. T., Pagliaro, E., & Tirnady, F. (1994). DNA typing in forensic science. I. Theory and background. Am J Forensic Med Pathol, 15(4), 269–282.

Lefebvre, L., Viville, S., Barton, S. C., Ishino, F., Keverne, E. B., & Surani, M. A. (1998). Abnormal maternal behaviour and growth retardation associated with loss of the imprinted gene Mest. Nat Genet, 20(2), 163–169.

Lenneberg, E. H. (1967). Biological foundations of language. New York: Wiley.

Lequin, M. H., & Barkovich, A. J. (1999). Current concepts of cerebral malformation syndromes. Curr Opin Pediatr, 11(6), 492–496.

Levinson, S. C., Kita, S., Haun, D. B., & Rasch, B. H. (2002). Returning the tables: language affects spatial reasoning. Cognition, 84(2), 155–188.

Levitt, P. (2000). Molecular determinants of regionalization of the forebrain and cerebral cortex. In M. S. Gazzaniga (ed.), The new cognitive neurosciences, 2d ed., pp. 23–32. Cambridge: MIT Press.

Leys, S. P., Mackie, G. O., & Meech, R. W. (1999). Impulse conduction in a sponge. J Exp Biol, 202 (Pt 9), 1139–1150.

Li, P., & Gleitman, L. (2002). Turning the tables: Language and spatial reasoning. Cognition, 83(3), 265–294.

Lieberman, P. (1984). The biology and evolution of language. Cambridge: Harvard University Press.

———. (2002). On the nature and evolution of the neural bases of human language. Am J Phys Anthropol, Suppl 35, 36–62.

Liittschwager, J. C., & Markman, E. M. (1994). Sixteen-and 24-month-olds' use of mutual exclusivity as a default assumption in second-label learning. Dev Psychol, 30(6), 955–968.

Linkenhoker, B. A., & Knudsen, E. I. (2002). Incremental training increases the plasticity of the auditory space map in adult barn owls. Nature, 419(6904), 293–296.

Liu, Q., Dwyer, N. D., & O'Leary, D. D. (2000). Differential expression of COUP-TFI, CHL1, and two novel genes in developing neocortex identified by differential display PCR. J Neurosci, 20(20), 7682–7690.

Loer, C. M., Steeves, J. D., & Goodman, C. S. (1983). Neuronal cell death in grasshopper embryos: variable patterns in different species, clutches, and clones. J Embryol Exp Morphol, 78, 169–182.

Logan, M., & Tabin, C. J. (1999). Role of Pitx1 upstream of Tbx4 in specification of hindlimb identity. Science, 283(5408), 1736–1739.

Lohmann, G., von Cramon, D. Y., & Steinmetz, H. (1999). Sulcal variability of twins. Cereb Cortex, 9(7), 754–763.

Lopez-Bendito, G., Shigemoto, R., Kulik, A., Paulsen, O., Fairen, A., & Lujan, R. (2002). Expression and distribution of metabotropic GABA receptor subtypes GABABR1 and GABABR2 during rat neocortical development. Eur J Neurosci, 15(11), 1766–1778.

Lykken, D. T. (1982). Research with twins: the concept of emergenesis. Psychophysiol, 19(4), 361–373.

Lykken, D. T., McGue, M., Tellegen, A., & Bouchard, T. J., Jr. (1992). Emergenesis. Genetic traits that may not run in families. Am Psychol, 47(12), 1565–1577.

Makalowski, W. (2000). Genomic scrap yard: how genomes utilize all that junk. Gene, 259(1–2), 61–67.

Mallamaci, A., Muzio, L., Chan, C. H., Parnavelas, J., & Boncinelli, E. (2000). Area identity shifts in the early cerebral cortex of Emx2-/- mutant mice. Nat Neurosci, 3(7), 679–686.

Mancama, D., & Kerwin, R. (2003). Role of pharmacogenomics in individualising treatment with SSRIs. CNS Drugs, 17(3), 143–151.

Manzanares, M., Wada, H., Itasaki, N., Trainor, P. A., Krumlauf, R., & Holland, P. W. (2000). Conservation and elaboration of Hox gene regulation during evolution of the vertebrate head. Nature, 408(6814), 854–857.

Marcus, G. F. (2000). Pa bi ku and ga ti ga: two mechanisms children could use to learn about language and the world. Current Directions in Psychol Sci, 9, 145–147.

____. (2001a). The algebraic mind: integrating connectionism and cognitive science. Cambridge: MIT Press.

____. (2001b). Plasticity and nativism: towards a resolution of an apparent paradox. In S. Wermter, J. Austin, & D. Willshaw (eds.), Emergent neural computational architectures based on neuroscience, pp. 368–382. New York: Springer-Verlag.

Marcus, G. F., & Fisher, S. E. (2003). FOXP2 in focus: what can genes tell us about speech and language? Trends Cogn Sci, 7, 257–262.

Marcus, G. F., Vijayan, S., Bandi Rao, S., & Vishton, P. M. (1999). Rule learning in 7-month-old infants. Science, 283, 77–80.

Margulies, E. H., Kardia, S. L., & Innis, J. W. (2001). A comparative molecular analysis of developing mouse forelimbs and hindlimbs using serial analysis of gene expression (sage). Genome Res, 11(10), 1686–1698.

Marin, O., & Rubenstein, J. L. (2001). A long, remarkable journey: tangential migration in the telencephalon. Nat Rev Neurosci, 2(11), 780–790.

____. (2003). Cell migration in the forebrain. Annu Rev Neurosci, 26, 441–483.

Marin, O., Yaron, A., Bagri, A., Tessier-Lavigne, M., & Rubenstein, J. L. (2001). Sorting of striatal and cortical interneurons regulated by semaphorin-neuropilin interactions. Science, 293(5531), 872–875.

Markman, E. M. (1989). Categorization and naming in children: problems of induction. Cambridge: MIT Press.

Marks, P., Iyer, G., Cui, Y., & Merchant, J. L. (1996). Fos is required for EGF stimulation of the gastrin promoter. Am J Physiol, 271(6 Pt 1), G942–948.

Marler, P. (1984). Song learning: innate species differences in the learning process. In P. Marler & H. Terrace (eds.), The biology of learning, pp. 289–309. Berlin: Springer-Verlag.

____. (1991). The instinct to learn. In S. Carey & R. Gelman (eds.), The epigenesis of mind: essays on biology and cognition, pp. 37–66. Hillsdale, N.J.: Lawrence Erlbaum Associates.

Martin, A., Haxby, J. V., Lalonde, F. M., Wiggs, C. L., & Ungerleider, L. G. (1995). Discrete cortical regions associated with knowledge of color and knowledge of action. Science, 270(5233), 102–105.

Martin, A., Wiggs, C. L., Ungerleider, L. G., & Haxby, J. V. (1996). Neural correlates of category-specific knowledge. Nature, 379(6566), 649–652.

Martin, S. (1998). Pure drivel. New York: Hyperion.

Martin, S. J., Grimwood, P. D., & Morris, R. G. (2000). Synaptic plasticity and memory: an evaluation of the hypothesis. Annu Rev Neurosci, 23, 649–711.

Maruishi, M., Mano, Y., Sasaki, T., Shinmyo, N., Sato, H., & Ogawa, T. (2001). Cerebral palsy in adults: independent effects of muscle strength and muscle tone. Arch Phys Med Rehabil, 82(5), 637–641.

Masland, R. H. (2001). Neuronal diversity in the retina. Curr Opin Neurobiol, 11(4), 431–436.

Mastronarde, D. N. (1983). Correlated firing of cat retinal ganglion cells. I. Spontaneously active inputs to X- and Y-cells. J Neurophysiol, 49(2), 303–324.

Mattson, M. P. (2002). Brain evolution and lifespan regulation: conservation of signal transduction pathways that regulate energy metabolism. Mech Ageing Dev, 123(8), 947–953.

Mayford, M., Bach, M. E., Huang, Y. Y., Wang, L., Hawkins, R. D., & Kandel, E. R. (1996). Control of memory formation through regulated expression of a CaMKII transgene. Science, 274(5293), 1678–1683.

Maynard Smith, J., & Szathmáry, E. (1995). The major transitions in evolution. Oxford: W. H. Freeman.

McIlwain, H., & Bachelard, H. S. (1985). Biochemistry and the central nervous system, 5th ed. Edinburgh and New York: Churchill Livingstone.

Medawar, P. B. (1981). Stretch genes. The New York Review of Books, 28 (July 16), 45–48.

Meister, M., Wong, R. O., Baylor, D. A., & Shatz, C. J. (1991). Synchronous bursts of action potentials in ganglion cells of the developing mammalian retina. Science, 252(5008), 939–943.

Mello, C. V., Vicario, D. S., & Clayton, D. F. (1992). Song presentation induces gene expression in the songbird forebrain. Proc Natl Acad Sci USA, 89(15), 6818–6822.

Meltzoff, A. N., & Moore, M. K. (1977). Imitation of facial and manual gestures by human neonate. Science, 198, 75–78.

Menand, L. (2002). What comes naturally: does evolution explain who we are? The New Yorker (November 25), 96–101.

Merzenich, M. M., Nelson, R. J., Stryker, M. P., Cynader, M. S., Schoppmann, A., & Zook, J. M. (1984). Somatosensory cortical map changes following digit amputation in adult monkeys. J Comp Neurol, 224(4), 591–605.

Metin, C., Denizot, J. P., & Ropert, N. (2000). Intermediate zone cells express calcium-permeable AMPA receptors and establish close contact with growing axons. J Neurosci, 20(2), 696–708.

Mey, J. (2001). Retinoic acid as a regulator of cytokine signaling after nerve injury. Z Naturforsch [C], 56(3–4), 163–176.

Miller, G. F. (2000). The mating mind: how sexual choice shaped the evolution of human nature. New York: Doubleday.

Mithen, S. J. (1996). The prehistory of the mind: a search for the origins of art, religion, and science. London: Thames and Hudson.

Miyashita-Lin, E. M., Hevner, R., Wassarman, K. M., Martinez, S., & Rubenstein, J. L. (1999). Early neocortical regionalization in the absence of thalamic innervation. Science, 285(5429), 906–909.

Modrek, B., Resch, A., Grasso, C., & Lee, C. (2001). Genome-wide detection of alternative splicing in expressed sequences of human genes. Nucleic Acids Res, 29(13), 2850–2859.

Mojzsis, S. J., Arrhenius, G., McKeegan, K. D., Harrison, T. M., Nutman, A. P., & Friend, C. R. (1996). Evidence for life on earth before 3,800 million years ago. Nature, 384(6604), 55–59.

Molnar, Z., Lopez-Bendito, G., Small, J., Partridge, L. D., Blakemore, C., & Wilson, M. C. (2002). Normal development of embryonic thalamocortical connectivity in the absence of evoked synaptic activity. J Neurosci, 22(23), 10313–10323.

Mombaerts, P. (1999). Molecular biology of odorant receptors in vertebrates. Annu Rev Neurosci, 22, 487–509.

Momose-Sato, Y., Miyakawa, N., Mochida, H., Sasaki, S., & Sato, K. (2003). Optical analysis of depolarization waves in the embryonic brain: a dual network of gap junctions and chemical synapses. J Neurophysiol, 89(1), 600–614.

Montgomery, S. A. (1999). New developments in the treatment of depression. J Clin Psych, 60 Suppl 14, 10–15; discussion 31–35.

Moon, C. M., & Fifer, W. P. (2000). Evidence of transnatal auditory learning. J Perinatol, 20(8 Pt 2), S37–44.

Moore, S., & Simon, J. L. (2000). It's getting better all the time: 100 greatest trends of the 20th century. Washington, D.C.: Cato Institute.

Morange, M. (1998). A history of molecular biology. Cambridge: Harvard University Press.

____. (2001). The misunderstood gene. Cambridge: Harvard University Press.

Morgan, E. (1995). The descent of the child: human evolution from a new perspective. New York: Oxford University Press.

Morimoto, T., Miyoshi, T., Fujikado, T., Tano, Y., & Fukuda, Y. (2002). Electrical stimulation enhances the survival of axotomized retinal ganglion cells in vivo. Neuroreport, 13(2), 227–230.

Morrison, G. E., & van der Kooy, D. (2001). A mutation in the AMPA-type glutamate receptor, glr–1, blocks olfactory associative and nonassociative learning in Caenorhabditis elegans. Behav Neurosci, 115(3), 640–649.

Morrison, G. E., Wen, J. Y., Runciman, S., & van der Kooy, D. (1999). Olfactory associative learning in Caenorhabditis elegans is impaired in lrn–1 and lrn–2 mutants. Behav Neurosci, 113(2), 358–367.

Morrongiello, B. A., Fenwick, K. D., & Chance, G. (1998). Crossmodal learning in newborn infants: inferences about properties of auditory-visual events. Infant Behav & Dev, 21(4), 543–553.

Munakata, Y., McClelland, J. L., Johnson, M. H., & Siegler, R. S. (1997). Rethinking infant knowledge: toward an adaptive process account of successes and failures in object permanence tasks. Psychol Rev, 10(4), 686–713.

Murphy, D. L., Li, Q., Engel, S., Wichems, C., Andrews, A., Lesch, K. P., et al. (2001). Genetic perspectives on the serotonin transporter. Brain Res Bull, 56(5), 487–494.

Nadeau, S. E., & Crosson, B. (1997). Subcortical aphasia. Brain Lang, 58(3), 355–402; discussion 418–423.

Nakagawa, Y., & O'Leary, D. D. (2001). Combinatorial expression patterns of LIM-homeodomain and other regulatory genes parcellate developing thalamus. J Neurosci, 21(8), 2711–2725.

Nasar, S. (1998). A beautiful mind: a biography of John Forbes Nash, Jr., winner of the Nobel Prize in economics, 1994. New York: Simon & Schuster.

Nazzi, T., Bertoncini, J., & Mehler, J. (1998). Language discrimination by newborns: towards an understanding of the role of rhythm. J Exp Psychol: Hum Percept & Perform, 24, 1–11.

Nazzi, T., Floccia, C., & Bertoncini, J. (1998). Discrimination of pitch contours by neonates. Infant Behav & Dev, 21(4), 779–784.

Nedivi, E., Hevroni, D., Naot, D., Israeli, D., & Citri, Y. (1993). Numerous candidate plasticity-related genes revealed by differential cDNA cloning. Nature, 363(6431), 718–722.

Nelkin, D. (2001). Molecular metaphors: the gene in popular discourse. Nat Rev Genet, 2(7), 555–559.

Nesse, R. M., & Williams, G. C. (1994). Why we get sick: the new science of Darwinian medicine. New York: Times Books.

Neville, H. J., & Lawson, D. (1987). Attention to central and peripheral visual space in a movement detection task: an event-related potential and behavioral study. II. Congenitally deaf adults. Brain Res, 405(2), 268–283.

Nottebohm, F., Stokes, T. M., & Leonard, C. M. (1976). Central control of song in the canary, Serinus canarius. J Comp Neurol, 165(4), 457–486.

O'Donovan, C., Apweiler, R., & Bairoch, A. (2001). The human proteomics initiative (HPI). Trends Biotechnol, 19(5), 178–181.

O'Leary, D. D., & Nakagawa, Y. (2002). Patterning centers, regulatory genes and extrinsic mechanisms controlling arealization of the neocortex. Curr Opin Neurobiol, 12(1), 14–25.

O'Leary, D. D., & Stanfield, B. B. (1989). Selective elimination of axons extended by developing cortical neurons is dependent on regional locale: experiments using fetal cortical transplants. J Neurosci, 9, 2230–2246.

O'Rahilly, R., Müller, F., & Streeter, G. L. (1987). Developmental stages in human embryos: including a revision of Streeter's "Horizons" and a survey of the Carnegie Collection. Washington, D.C.: Carnegie Institution of Washington.

Ochs, E., & Schieffelin, B. B. (1984). Language acquisition and socialization: three developmental stories and their implications. In A. Shweder Richard & A. LeVine Robert (eds.), Culture theory: essays on mind, self, and emotion, pp. 276–320. Cambridge: Cambridge University Press.

Oetting, W., & Bennett, D. (2003). Mouse coat color genes. International Federation of Pigment Cell Societies. Available at http://www.cbc.umn.edu/ifpcs/micemut.htm.

Ohno, S. (1970). Evolution by gene duplication. Berlin and New York: Springer-Verlag.

Ojemann, G. A. (1993). Functional mapping of cortical language areas in adults: intraoperative approaches. Adv Neurol, 63, 155–163.

Oksenberg, J. R., Barcellos, L. F., & Hauser, S. L. (1999). Genetic aspects of multiple sclerosis. Semin Neurol, 19(3), 281–288.

Olby, R. C. (1994). The path to the double helix: the discovery of DNA. New York: Dover Publications.

Onishi, K., & Baillargeon, R. (2002). Fifteen-month-old infants' understanding of false belief. Biennial International Conference on Infant Studies, Toronto, Canada.

Oram, M. W., & Perrett, D. I. (1992). Time course of neural responses discriminating different views of the face and head. J Neurophysiol, 68(1), 70–84.

Ortells, M. O., & Lunt, G. G. (1995). Evolutionary history of the ligand-gated ion-channel superfamily of receptors. Trends Neurosci, 18(3), 121–127.

Pagliarulo, V., Datar, R. H., & Cote, R. J. (2002). Role of genetic and expression profiling in pharmacogenomics: the changing face of patient management. Curr Issues Mol Biol, 4(4), 101–110.

Pascalis, O., de Haan, M., & Nelson, C. A. (2002). Is face processing species-specific during the first year of life? Science, 296(5571), 1321–1323.

Patthy, L. (2003). Modular assembly of genes and the evolution of new functions. Genetica, 118(2–3), 217–31.

Paul, D. B., & Blumenthal, A. L. (1989). On the trail of Little Albert. Psychol Rec, 39(4), 547–553.

Pena De Ortiz, S., & Arshavsky, Y. (2001). DNA recombination as a possible mechanism in declarative memory: a hypothesis. J Neurosci Res, 63(1), 72–81.

Penn, A. A., & Shatz, C. J. (1999). Brain waves and brain wiring: the role of endogenous and sensory-driven neural activity in development. Pediatr Res, 45(4 Pt 1), 447–458.

Pennartz, C. M., Uylings, H. B., Barnes, C. A., & McNaughton, B. L. (2002). Memory reactivation and consolidation during sleep: from cellular mechanisms to human performance. Prog Brain Res, 138, 143–166.

Pennington, B. F., Filipek, P. A., Lefly, D., Chhabildas, N., Kennedy, D. N., Simon, J. H., et al. (2000). A twin MRI study of size variations in the human brain. J Cogn Neurosci, 12(1), 223–232.

Pentland, A. (1997). Content-based indexing of images and video. Philos Trans R Soc Lond B Biol Sci, 352(1358), 1283–1290.

Phan, K. L., Wager, T., Taylor, S. F., & Liberzon, I. (2002). Functional neuroanatomy of emotion: a meta-analysis of emotion activation studies in PET and fMRI. Neuroimage, 16(2), 331–348.

Piaget, J. (1954). The construction of reality in the child. New York: Basic Books.

Pinker, S. (1994). The language instinct. New York: Morrow.

___. (1997). How the mind works. New York: Norton.

___. (2002). The blank slate. New York: Viking Penguin.

Pinker, S., & Bloom, P. (1990). Natural language and natural selection. Behav & Brain Sci, 13, 707–784.

Pizzorusso, T., Medini, P., Berardi, N., Chierzi, S., Fawcett, J. W., & Maffei, L. (2002). Reactivation of ocular dominance plasticity in the adult visual cortex. Science, 298(5596), 1248–1251.

Plomin, R. (1997). Behavioral genetics, 3d ed. New York: W. H. Freeman.

Plomin, R., & Crabbe, J. (2000). DNA. Psychol Bull, 126(6), 806–828.

Plomin, R., DeFries, J. C., McClearn, G. E., & McGuffin, P. (2001). Behavioral genetics, 4th ed. New York: Worth.

Plomin, R., & McGuffin, P. (2003). Psychopathology in the postgenomic era. Annu Rev Psychol, 54(1), 205–228.

Posthuma, D., De Geus, E. J., Baare, W. F., Pol, H. E., Kahn, R. S., & Boomsma, D. I. (2002). The association between brain volume and intelligence is of genetic origin. Nat Neurosci, 5(2), 83–84.

Postle, B. R., & Corkin, S. (1998). Impaired word-stem completion priming but intact perceptual identification priming with novel words: evidence from the amnesic patient H. M. Neuropsychologia, 36(5), 421–440.

Povinelli, D. J. (2000). Folk physics for apes: the chimpanzee's theory of how the world works. Oxford and New York: Oxford University Press.

Profet, M. (1992). Pregnancy sickness as adaptation: a deterrent to maternal ingestion of teratogens. In J. Barkow, J. Tooby, & L. Cosmides (eds.), The adapted mind: evolutionary psychology and the generation of culture, pp. 327–365. Oxford: Oxford University Press.

Pulvermuller, F. (2002). A brain perspective on language mechanisms: from discrete neuronal ensembles to serial order. Prog Neurobiol, 67(2), 85–111.

Purves, W. K., Sadava, D., Orians, G. H., & Heller, H. C. (2001). Life, the science of biology, 6th ed. Sunderland, Mass.: Sinauer Associates.

Quartz, S. R., & Sejnowski, T. J. (1997). The neural basis of cognitive development: a constructivist manifesto. Behav & Brain Sci, 20, 537–556; discussion 556–596.

Rajagopalan, S., Vivancos, V., Nicolas, E., & Dickson, B. J. (2000). Selecting a longitudinal pathway: Robo receptors specify the lateral position of axons in the Drosophila CNS. Cell, 103(7), 1033–1045.

Rakic, P. (1972). Mode of cell migration to the superficial layers of fetal monkey neocortex. J Comp Neurol, 145(1), 61–83.

___. (1998). Young neurons for old brains? Nature Neurosci, 1(8), 645–647.

Ramos, J. M. (2000). Long-term spatial memory in rats with hippocampal lesions. Eur J Neurosci, 12(9), 3375–3384.

Rampon, C., Jiang, C. H., Dong, H., Tang, Y. P., Lockhart, D. J., Schultz, P. G., et al. (2000). Effects of environmental enrichment on gene expression in the brain. Proc Natl Acad Sci USA, 97(23), 12880–12884.

Ramus, F., Hauser, M. D., Miller, C., Morris, D., & Mehler, J. (2000). Language discrimination by human newborns and by cotton-top tamarin monkeys. Science, 288(5464), 349–351.

Ramus, F., Rosen, S., Dakin, S. C., Day, B. L., Castellote, J. M., White, S., et al. (2003). Theories of developmental dyslexia: insights from a multiple case study of dyslexic adults. Brain, 126(Pt 4), 841–865.

Rankin, C. H. (2002). From gene to identified neuron to behaviour in Caenorhabditis elegans. Nat Rev Genet, 3(8), 622–630.

Rauschecker, J. P. (1995). Compensatory plasticity and sensory substitution in the cerebral cortex. Trends Neurosci, 18(1), 36–43.

Rebillard, G., Carlier, E., Rebillard, M., & Pujol, R. (1977). Enhancement of visual responses on the primary auditory cortex of the cat after an early destruction of cochlear receptors. Brain Res, 129(1), 162–164.

Redies, C. (2000). Cadherins in the central nervous system. Prog Neurobiol, 61(6), 611–648.

Regolin, L., Tommasi, L., & Vallortigara, G. (2000). Visual perception of biological motion in newly hatched chicks as revealed by an imprinting procedure. Anim Cogn, 3(1), 53–60.

Regolin, L., Vallortigara, G., & Zanforlin, M. (1995). Object and spatial representations in detour problems by chicks. Anim Behav, 49, 195–199.

Reh, T. A., & Constantine-Paton, M. (1985). Eye-specific segregation requires neural activity in three-eyed Rana pipiens. J Neurosci, 5(5), 1132–1143.

Renner, M. (1960). Contribution of the honey bee to the study of time sense and astronomical orientation. Cold Spring Harbor Symposium on Quantitative Biology, 25, 361–367.

Rensberger, B. (1996). Life itself: exploring the realm of the living cell. New York: Oxford University Press.

Restak, R. M. (1979). The brain: the last frontier. Garden City, N.Y.: Doubleday.

Rice, D. S., & Curran, T. (2001). Role of the reelin signaling pathway in central nervous system development. Annu Rev Neurosci, 24, 1005–1039.

Richardson, M. K., Hanken, J., Selwood, L., Wright, G. M., Richards, R. J., Pieau, C., et al. (1998). Haeckel, embryos, and evolution. Science, 280(5366), 983, 985–986.

Richardson, W. D., Pringle, N. P., Yu, W. P., & Hall, A. C. (1997). Origins of spinal cord oligodendrocytes: possible developmental and evolutionary relationships with motor neurons. Dev Neurosci, 19(1), 58–68.

Richerson, P. J., & Boyd, R. (Forthcoming). The nature of cultures.

Rilling, J. K., & Insel, T. R. (1999). Differential expansion of neural projection systems in primate brain evolution. Neuroreport, 10(7), 1453–1459.

Rivera, S. M., Wakeley, A., & Langer, J. (1999). The drawbridge phenomenon: representational reasoning or perceptual preference. Dev Psychol, 35, 427–435.

Rizzolatti, G., Fadiga, L., Gallese, V., & Fogassi, L. (1996). Premotor cortex and the recognition of motor actions. Cogn Brain Res, 3(2), 131–141.

Roberson, D., Davies, I., & Davidoff, J. (2000). Color categories are not universal: replications and new evidence from a stone-age culture. J Exp Psychol Gen, 129(3), 369–398.

Rose, S. P. R. (1973). The conscious brain. New York: Knopf.

____. (2000). God's organism? The chick as a model system for memory studies. Learn Mem, 7(1), 1–17.

Rosen, K. M., McCormack, M. A., Villa-Komaroff, L., & Mower, G. D. (1992). Brief visual experience induces immediate early gene expression in the cat visual cortex. Proc Natl Acad Sci USA, 89(12), 5437–5441.

Ross, M. E., & Walsh, C. A. (2001). Human brain malformations and their lessons for neuronal migration. Annu Rev Neurosci, 24, 1041–1070.

Rossi, F., & Cattaneo, E. (2002). Opinion: neural stem cell therapy for neurological diseases: dreams and reality. Nat Rev Neurosci, 3(5), 401–409.

Roth, K. A., & D'sa, C. (2001). Apoptosis and brain development. Ment Retard Dev Disabil Res Rev, 7(4), 261–266.

Rowe, D. C. (1994). The limits of family influence: genes, experience, and behavior. New York: Guilford.

Ruiz-Trillo, I., Riutort, M., Littlewood, D. T., Herniou, E. A., & Baguna, J. (1999). Acoel flatworms: earliest extant bilaterian Metazoans, not members of Platyhelminthes. Science, 283(5409), 1919–1923.

Sachs, B. D. (1988). The development of grooming and its expression in adult animals. Ann NY Acad Sci, 525, 1–17.

Sadato, N., Pascual-Leone, A., Grafman, J., Ibanez, V., Deiber, M. P., Dold, G., et al. (1996). Activation of the primary visual cortex by Braille reading in blind subjects. Nature, 380(6574), 526–528.

Saffran, J., Aslin, R., & Newport, E. (1996). Statistical learning by 8-month-old infants. Science, 274, 1926–1928.

Sagan, C., & Druyan, A. (1992). Shadows of forgotten ancestors: a search for who we are. New York: Random House.

Sanes, D. H., Reh, T. A., & Harris, W. A. (2000). Development of the nervous system. San Diego and London: Academic.

Sanes, J. R., & Lichtman, J. W. (1999). Can molecules explain long-term potentiation? Nat Neurosci, 2(7), 597–604.

Sapolsky, R. M. (2003). Gene therapy for psychiatric disorders. Am J Psych, 160(2), 208–220.

Sarnat, H. B., & Netsky, M. G. (1985). The brain of the planarian as the ancestor of the human brain. Can J Neurol Sci, 12(4), 296–302.

Saunders, J. W. (1982). Developmental biology: patterns, problems, and principles. New York: Macmillan.

Saunders, J. W., Gasseling, M. T., & Cairns, J. M. (1959). The differentiation of prospective thigh mesoderm grafted beneath the apical ectodermal ridge of the wing bud in the chick embryo. Dev Biol, 1, 281–301.

Savage-Rumbaugh, E. S., Murphy, J., Sevcik, R. A., Brakke, K. E., Williams, S. L., & Rumbaugh, D. M. (1993). Language comprehension in ape and child. Monogr Soc Res Child Dev, 58(3–4), 1–222.

Sayre, A. (1975). Rosalind Franklin and DNA. New York: Norton.

Scamvougeras, A., Kigar, D. L., Jones, D., Weinberger, D. R., & Witelson, S. F. (2003). Size of the human corpus callosum is genetically determined: an MRI study in mono and dizygotic twins. Neurosci Lett, 338(2), 91–94.

Schacter, D. L. (1996). Searching for memory: the brain, the mind, and the past. New York: Basic Books.

Schad, W. (1993). Heterochronical patterns of evolution in the transitional stages of vertebrate classes. Acta Biotheor, 41(4), 383–389.

Scheller, R. H., & Axel, R. (1984). How genes control an innate behavior. Sci Amer(March), 54–62.

Schlaug, G. (2001). The brain of musicians: a model for functional and structural adaptation. Ann NY Acad Sci, 930, 281–299.

Schmidt, J. T., & Eisele, L. E. (1985). Stroboscopic illumination and dark rearing block the sharpening of the regenerated retinotectal map in goldfish. Neurosci, 14(2), 535–546.

Schmucker, D., Clemens, J. C., Shu, H., Worby, C. A., Xiao, J., Muda, M., et al. (2000). Drosophila Dscam is an axon guidance receptor exhibiting extraordinary molecular diversity. Cell, 101(6), 671–684.

Scoville, W. B., & Milner, B. (1957). Loss of recent memory after bilateral hippocampal lesions. J Neurol Neurosurg Psych, 20, 11–21.

Seeman, P., & Madras, B. (2002). Methylphenidate elevates resting dopamine which lowers the impulse-triggered release of dopamine: a hypothesis. Behav Brain Res, 130(1–2), 79–83.

Seidl, R., Cairns, N., & Lubec, G. (2001). The brain in Down syndrome. J Neural Transm Suppl(61), 247–261.

Semendeferi, K., & Damasio, H. (2000). The brain and its main anatomical subdivisions in living hominoids using magnetic resonance imaging. J Hum Evol, 38(2), 317–332.

Sestan, N., Rakic, P., & Donoghue, M. J. (2001). Independent parcellation of the embryonic visual cortex and thalamus revealed by combinatorial Eph/ephrin gene expression. Curr Biol, 11(1), 39–43.

Seyfarth, R. M., Cheney, D. L., & Marler, P. (1980). Monkey responses to three different alarm calls: evidence of predator classification and semantic communication. Science, 210(4471), 801–803.

Shapleske, J., Rossell, S. L., Woodruff, P. W., & David, A. S. (1999). The planum temporale: a systematic, quantitative review of its structural, functional and clinical significance. Brain Res Rev, 29(1), 26–49.

Sharma, K., Leonard, A. E., Lettieri, K., & Pfaff, S. L. (2000). Genetic and epigenetic mechanisms contribute to motor neuron pathfinding. Nature, 406(6795), 515–519.

Shaywitz, S. E., Fletcher, J. M., Holahan, J. M., Shneider, A. E., Marchione, K. E., Stuebing, K. K., et al. (1999). Persistence of dyslexia: the Connecticut longitudinal study at adolescence. Pediatrics, 104(6), 1351–1359.

Shell, E. R. (2002). The hungry gene: the science of fat and the future of thin. New York: Atlantic Monthly Press.

Sherrington, R., Brynjolfsson, J., Petursson, H., Potter, M., Dudleston, K., Barraclough, B., et al. (1988). Localization of a susceptibility locus for schizophrenia on chromosome 5. Nature, 336(6195), 164–167.

Sherry, D. F., Jacobs, L. F., & Gaulin, S. J. (1992). Spatial memory and adaptive specialization of the hippocampus. Trends Neurosci, 15(8), 298–303.

Shors, T. J., & Matzel, L. D. (1997). Long-term potentiation: what's learning got to do with it? Behav Brain Sci, 20(4), 597–614; discussion 614–655.

Shu, D., Luo, H., Morris, S., Zhang, X., Hu, S., Chen, L., et al. (1992). Lower Cambrian vertebrates from South China. Nature, 402(6757), 42–46.

Shu, W., Yang, H., Zhang, L., Lu, M. M., & Morrisey, E. E. (2001). Characterization of a new subfamily of winged-helix/forkhead (Fox) genes that are expressed in the lung and act as transcriptional repressors. J Biol Chem, 276(29), 27488–27497.

Sillaber, I., Rammes, G., Zimmermann, S., Mahal, B., Zieglgansberger, W., Wurst, W., et al. (2002). Enhanced and delayed stress-induced alcohol drinking in mice lacking functional CRH1 receptors. Science, 296(5569), 931–933.

Silva, A. J., Paylor, R., Wehner, J. M., & Tonegawa, S. (1992). Impaired spatial learning in alpha-calcium-calmodulin kinase II mutant mice. Science, 257(5067), 206–211.

Silva, A. J., Stevens, C. F., Tonegawa, S., & Wang, Y. (1992). Deficient hippocampal long-term potentiation in alpha-calcium-calmodulin kinase II mutant mice. Science, 257(5067), 201–206.

Simeone, A., Puelles, E., & Acampora, D. (2002). The Otx family. Curr Opin Genet Dev, 12(4), 409–415.

Simpson, J. H., Bland, K. S., Fetter, R. D., & Goodman, C. S. (2000). Short-range and long-range guidance by slit and its Robo receptors: a combinatorial code of Robo receptors controls lateral position. Cell, 103(7), 1019–1032.

Skeath, J. B., & Thor, S. (2003). Genetic control of Drosophila nerve cord development. Curr Opin Neurobiol, 13(1), 8–15.

Skoyles, J. (1999). Human evolution expanded brains to increase expertise capacity, not IQ. Psycoloquy, 10(2).

Smith, L. B., Thelen, E., Titzer, R., & McLin, D. (1999). Knowing in the context of acting: the task dynamics of the A-not-B error. Psychol Rev, 106(2), 235–260.

Smith, V. A., King, A. P., & West, M. J. (2000). A role of her own: female cowbirds, Molothrus ater, influence the development and outcome of song learning. Anim Behav, 60(5), 599–609.

Sokolowski, M. B. (1998). Genes for normal behavioral variation: recent clues from flies and worms. Neuron, 21(3), 463–466.

Song, B., Zhao, M., Forrester, J. V., & McCaig, C. D. (2002). Electrical cues regulate the orientation and frequency of cell division and the rate of wound healing in vivo. Proc Natl Acad Sci USA, 99(21), 13577–13582.

Song, H. J., Billeter, J. C., Reynaud, E., Carlo, T., Spana, E. P., Perrimon, N., et al. (2002). The fruitless gene is required for the proper formation of axonal tracts in the embryonic central nervous system of Drosophila. Genetics, 162(4), 1703–1724.

Sperry, R. W. (1961). Cerebral organization and behavior. Science, 133, 1749–1757.

Spock, B. (1957). Baby and child care, 2d ed. New York: Pocket Books.

Stanfield, B. B., & O'Leary, D. D. (1985). Fetal occipital cortical neurons transplanted to the rostral cortex can extend and maintain a pyramidal tract axon. Nature, 313, 135–137.

Stein, J. (2001). The magnocellular theory of developmental dyslexia. Dyslexia, 7(1), 12–36.

Stellwagen, D., & Shatz, C. J. (2002). An instructive role for retinal waves in the development of retinogeniculate connectivity. Neuron, 33(3), 357–367.

Stephan, H., Frahm, H., & Baron, G. (1981). New and revised data on volumes of brain structures in insectivores and primates. Folia Primatol (Basel), 35(1), 1–29.

Stock, G. (2002). Redesigning humans: our inevitable genetic future. Boston: Houghton Mifflin.

Stromswold, K., Caplan, D., Alpert, N., & Rauch, S. (1996). Localization of syntactic comprehension by positron emission tomography. Brain Lang, 52(3), 452–473.

Stuhmer, T., Anderson, S. A., Ekker, M., & Rubenstein, J. L. (2002). Ectopic expression of the Dlx genes induces glutamic acid decarboxylase and Dlx expression. Development, 129(1), 245–252.

Sturtevant, A. H. (1913). The linear arrangement of six sex-linked factors in Drosophila, as shown by their mode of association. J Exp Zool, 14, 43–59.

Sur, M., & Leamey, C. A. (2001). Development and plasticity of cortical areas and networks. Nat Rev Neurosci, 2(4), 251–262.

Sur, M., Pallas, S. L., & Roe, A. W. (1990). Cross-model plasticity in cortical development: differentiation and specification of sensory neocortex. Trends Neurosci, 13, 227–233.

Takeuchi, J. K., Koshiba-Takeuchi, K., Matsumoto, K., Vogel-Hopker, A., Naitoh-Matsuo, M., Ogura, K., et al. (1999). Tbx5 and Tbx4 genes determine the wing/leg identity of limb buds. Nature, 398(6730), 810–814.

Tanford, C., & Reynolds, J. A. (2001). Nature's robots: a history of proteins. Oxford and New York: Oxford University Press.

Tang, Y. P., Shimizu, E., Dube, G. R., Rampon, C., Kerchner, G. A., Zhuo, M., et al. (1999). Genetic enhancement of learning and memory in mice. Nature, 401(6748), 63–69.

Tavare, S., Marshall, C. R., Will, O., Soligo, C., & Martin, R. D. (2002). Using the fossil record to estimate the age of the last common ancestor of extant primates. Nature, 416(6882), 726–729.

Tecott, L. H. (2003). The genes and brains of mice and men. Am J Psych, 160(4), 646–656.

Temple, E., Deutsch, G. K., Poldrack, R. A., Miller, S. L., Tallal, P., Merzenich, M. M., et al. (2003). Neural deficits in children with dyslexia ameliorated by behavioral remediation: evidence from functional MRI. Proc Natl Acad Sci USA, 100(5), 2860–2865.

Thompson, P. M., Cannon, T. D., Narr, K. L., van Erp, T., Poutanen, V. P., Huttunen, M., et al. (2001). Genetic influences on brain structure. Nat Neurosci, 4(12), 1253–1258.

Thulborn, K. R., Carpenter, P. A., & Just, M. A. (1999). Plasticity of language-related brain function during recovery from stroke. Stroke, 30(4), 749–754.

Tiihonen, J., Kuikka, J., Kupila, J., Partanen, K., Vainio, P., Airaksinen, J., et al. (1994). Increase in cerebral blood flow of right prefrontal cortex in man during orgasm. Neurosci Lett, 170(2), 241–243.

Tole, S., Goudreau, G., Assimacopoulos, S., & Grove, E. A. (2000). Emx2 is required for growth of the hippocampus but not for hippocampal field specification. J Neurosci, 20(7), 2618–2625.

Tomasello, M. (1999). The cultural origins of human cognition. Cambridge: Harvard University Press.

Tomasello, M., Call, J., & Hare, B. (2003). Chimpanzees understand psychological states—the question is which ones and to what extent. Trends Cogn Sci, 7(4), 153–156.

Tsai, Y. J., & Hoyme, H. E. (2002). Pharmacogenomics: the future of drug therapy. Clin Genet, 62(4), 257–264.

Tsien, J. Z., Huerta, P. T., & Tonegawa, S. (1996). The essential role of hippocampal CA1 NMDA receptor-dependent synaptic plasticity in spatial memory. Cell, 87(7), 1327–1338.

Tversky, A., & Gati, I. (1982). Similarity, separability, and the triangle inequality. Psychol Rev, 89(2), 123–154.

van der Lely, H. K., Rosen, S., & McClelland, A. (1998). Evidence for a grammar-specific deficit in children. Curr Biol, 8(23), 1253–1258.

van der Lely, H. K., & Stollwerck, L. (1996). A grammatical specific language impairment in children: an autosomal dominant inheritance? Brain Lang, 52(3), 484–504.

van Schaik, C. P., Ancrenaz, M., Borgen, G., Galdikas, B., Knott, C. D., Singleton, I., et al. (2003). Orangutan cultures and the evolution of material culture. Science, 299(5603), 102–105.

Vargha-Khadem, F., Gadian, D. G., Watkins, K. E., Connelly, A., Van Paesschen, W., & Mishkin, M. (1997). Differential effects of early hippocampal pathology on episodic and semantic memory. Science, 277(5324), 376–380. Erratum in Science 277(5329), Aug. 22, 1997, 1117.

Vargha-Khadem, F., Isaacs, E., & Muter, V. (1994). A review of cognitive outcome after unilateral lesions sustained during childhood. J Child Neurol, 9 Suppl 2, 67–73.

Vargha-Khadem, F., Watkins, K., Alcock, K., Fletcher, P., & Passingham, R. (1995). Praxic and nonverbal cognitive deficits in a large family with a genetically transmitted speech and language disorder. Proc Natl Acad Sci USA, 92(3), 930–933.

Venter, J. C., Adams, M. D., Myers, E. W., Li, P. W., Mural, R. J., Sutton, G. G., et al. (2001). The sequence of the human genome. Science, 291(5507), 1304–1351.

Verhaegen, M. J. M. (1988). Aquatic ape theory and speech origins: a hypothesis. Speculations in Science and Technology, 11, 165–171.

Verhage, M., Maia, A. S., Plomp, J. J., Brussaard, A. B., Heeroma, J. H., Vermeer, H., et al. (2000). Synaptic assembly of the brain in the absence of neurotransmitter secretion. Science, 287(5454), 864–869.

Vicari, S., Albertoni, A., Chilosi, A. M., Cipriani, P., Cioni, G., & Bates, E. (2000). Plasticity and reorganization during language development in children with early brain injury. Cortex, 36(1), 31–46.

Wallace, C. S., Withers, G. S., Weiler, I. J., George, J. M., Clayton, D. F., & Greenough, W. T. (1995). Correspondence between sites of NGFI-A induction and sites of morphological plasticity following exposure to environmental complexity. Mol Brain Res, 32(2), 211–220.

Walsh, G. (2002). Proteins: biochemistry and biotechnology. West Sussex, England, and New York: J. Wiley.

Warrington, J. A., Nair, A., Mahadevappa, M., & Tsyganskaya, M. (2000). Comparison of human adult and fetal expression and identification of 535 housekeeping/maintenance genes. Physiol Genomics, 2(3), 143–147.

Washbourne, P., Thompson, P. M., Carta, M., Costa, E. T., Mathews, J. R., Lopez-Bendito, G., et al. (2002). Genetic ablation of the t-SNARE SNAP–25 distinguishes mechanisms of neuroexocytosis. Nat Neurosci, 5(1), 19–26.

Waterston, R. H., Lindblad-Toh, K., Birney, E., Rogers, J., Abril, J. F., Agarwal, P., et al. (2002). Initial sequencing and comparative analysis of the mouse genome. Nature, 420(6915), 520–562.

Watkins, K. E., Dronkers, N. F., & Vargha-Khadem, F. (2002). Behavioural analysis of an inherited speech and language disorder: comparison with acquired aphasia. Brain, 125(Pt 3), 452–464.

Watkins, T. A., & Barres, B. A. (2002). Nerve regeneration: regrowth stumped by shared receptor. Curr Biol, 12(19), R654–656.

Watson, J. B. (1925). Behaviorism [microform]. New York: W. W. Norton.

Watson, J. D., & Crick, F. H. (1953). A structure for deoxyribose nucleic acid. Nature, 171, 737–738.

Webb, D. J., Parsons, J. T., & Horwitz, A. F. (2002). Adhesion assembly, disassembly and turnover in migrating cells—over and over and over again. Nat Cell Biol, 4(4), E97–100.

Webster, M. J., Ungerleider, L. G., & Bachevalier, J. (1995). Development and plasticity of the neural circuitry underlying visual recognition memory. Can J Physiol Pharmacol, 73(9), 1364–1371.

Wei, F., Wang, G. D., Kerchner, G. A., Kim, S. J., Xu, H. M., Chen, Z. F., et al. (2001). Genetic enhancement of inflammatory pain by forebrain NR2B over-expression. Nat Neurosci, 4(2), 164–169.

Weliky, M., & Katz, L. C. (1997). Disruption of orientation tuning in visual cortex by artificially correlated neuronal activity. Nature, 386(6626), 680–685.

Welker, E. (2000). Developmental plasticity: to preserve the individual or to create a new species? Novartis Found Symp, 228, 227–235; discussion 235–239.

Wells, M. J. (1966). Learning in the octopus. Symp Soc Exp Biol, 20, 477–507.

Wessler, I., Kirkpatrick, C. J., & Racke, K. (1999). The cholinergic "pitfall": acetyl-choline, a universal cell molecule in biological systems, including humans. Clin Exp Pharmacol Physiol, 26(3), 198–205.

White, J. G., Southgate, E., Thomson, J. N., & Brenner, S. (1986). The structure of the ventral nerve cord of Caenorhabditis elegans. Philos Trans R Soc Lond B Biol Sci (1165), 1–340.

Whiten, A., Goodall, J., McGrew, W. C., Nishida, T., Reynolds, V., Sugiyama, Y., et al. (1999). Cultures in chimpanzees. Nature, 399(6737), 682–685.

Whorf, B. L. (1975[1956]). The organization of reality. In S. Rogers (ed.), Children and language. New York: Oxford University Press.

Wickett, J. C., Vernon, P. A., & Lee, D. H. (2000). Relationships between factors of intelligence and brain volume. Personality & Individual Differences, 29(6), 1095–1122.

Wiesel, T. N., & Hubel, D. H. (1963). Single-cell responses in striate cortex of very young, visually inexperienced kittens. J Neurophysiol, 26, 1003–1017.

Wild, J. M. (1997). Neural pathways for the control of birdsong production. J Neurobiol, 33(5), 653–670.

Williams, N. A., & Holland, P. W. (2000). An amphioxus Emx homeobox gene reveals duplication during vertebrate evolution. Mol Biol Evol, 17(10), 1520–1528.

Wilmut, I., Schnieke, A. E., McWhir, J., Kind, A. J., & Campbell, K. H. (1997). Viable offspring derived from fetal and adult mammalian cells. Nature, 385(6619), 810–813.

Wilson, A. C., & Sarich, V. M. (1969). A molecular time scale for human evolution. Proc Natl Acad Sci USA, 63(4), 1088–1093.

Wimmer, H., & Perner, J. (1983). Beliefs about beliefs: representation and constraining function of wrong beliefs in young children's understanding of deception. Cognition, 13(1), 103–128.

Winsberg, B. G., & Comings, D. E. (1999). Association of the dopamine transporter gene (DAT1) with poor methylphenidate response. J Am Acad Child Adolesc Psych, 38(12), 1474–1477.

Wise, R. J., Scott, S. K., Blank, S. C., Mummery, C. J., Murphy, K., & Warburton, E. A. (2001). Separate neural subsystems within "Wernicke's area." Brain, 124(Pt 1), 83–95.

Wo, Z. G., & Oswald, R. E. (1995). Unraveling the modular design of glutamate-gated ion channels. Trends Neurosci, 18(4), 161–168.

Wong, R. O. (1999). Retinal waves and visual system development. Annu Rev Neurosci, 22, 29–47.

Wynn, K. (1992). Addition and subtraction by human infants. Nature, 358, 749–750.

Wynn, K. (2002). Do infants have numerical expectations or just perceptual preferences? Comment. Dev Sci, 5(2), 207–209.

Xue, H. (1998). Identification of major phylogenetic branches of inhibitory ligand-gated channel receptors. J Mol Evol, 47(3), 323–333.

Young, L. J., Nilsen, R., Waymire, K. G., MacGregor, G. R., & Insel, T. R. (1999). Increased affiliative response to vasopressin in mice expressing the V1a receptor from a monogamous vole. Nature, 400(6746), 766–768.

Younossi-Hartenstein, A., Jones, M., & Hartenstein, V. (2001). Embryonic development of the nervous system of the temnocephalid flatworm Craspedella pedum. J Comp Neurol, 434(1), 56–68.

Zatorre, R. J. (2001). Neural specializations for tonal processing. Ann NY Acad Sci, 930, 193–210.

Zhang, J., Webb, D. M., & Podlaha, O. (2002). Accelerated protein evolution and origins of human-specific features: Foxp2 as an example. Genetics, 162(4), 1825–1835.

Zhang, L. I., & Poo, M. M. (2001). Electrical activity and development of neural circuits. Nat Neurosci, 4 Suppl, 1207–1214.

INDEX

Note: page numbers in italics refer to glossary items; "f" following page numbers refers to figures. See also Name Index.

activity, brain, 191
adaptation, 169, 191
adult brain, structure of, 32f
ADD (Attention Deficit Disorder), 173
adrenaline, 115, 191
albinism, 53
alcoholism, 83, 180
alkaptonuria, 53, 191
alternative splicing, 82, 155, 191
Alzheimer's disease, 78
amino acids, 54, 56–57, 112, 187, 191
amnesia, 100
amphioxus, 114, 120, 191
amygdala, 43, 102–103, 191
amyotrophic lateral sclerosis (Lou Gehrig's
     disease), 53, 191
Angelman syndrome, 78
anemia, sickle-cell, 53, 58, 205
animal model, 133, 188, 191
animals
     culture and, 26–27, 128
     genomes in, 29, 79, 80
     innate abilities of, 20–21
     language and, 28, 29, 130–131
     learning abilities in, 22–23, 24–25
     See also specific animals
antinativists, 43, 45
ants, 66
anxiety, 77, 78, 79
Aplysia (sea slugs), 22, 81–82
apoptosis (cell death), 72–73, 75, 99, 194
approximation, successive, 48, 49
area TE, 38
association (in psychology), 22, 192
association (in genetics), 106, 184–186, 191
associative learning, 105, 192
The Astonishing Hypothesis, 1
atoms, 68

attention, 192
Attention Deficit Disorder (ADD), 173
auditory area, 122, 129, 192
auditory cortex, 37, 38, 122
autism, 192
     brain wiring and, 90
     lack of cure for, 44–45
     genes and, 78, 84, 180
     learning language and, 138
     reelin and, 121
autonomous agent, 50, 179, 192
Autonomous Agent Theory, 50, 59–60, 81
axon guidance molecules, 98, 150, 192
axons, 70f, 192
     function of, 69
     growth cones of, 91–93, 94
     guidance for, 94, 95, 96–97
     innate abilities of, 21
     myelin insulation and, 118
     retinal, 42–43
azimuth system, 24, 192

babies
     brain structure of, 12, 31–32, 32f
     cognitive development in, 4, 16–20
     intelligence and, 15–16
     language and, 28–30, 31–32
     learning abilities of, 25–26, 28–30, 31–32
     speech and, 27–28
baboons, 27
barrel fields, 74, 96, 169, 192
basal ganglia, 119, 192
base, DNA, 192
Basic Local Alignment Search Tool (BLAST),
     189
Bauplan, 166
A Beautiful Mind, 180
behavior disorders, 188

NAME INDEX

Note: "n" following page numbers refers to endnotes. Number following the "n", and in parentheses, refers to specific note.